CHEMICAL ENGINEERING

VOLUME 4

Solutions to the Problems in Chemical Engineering Volume 1

D1194263

CHEMICAL ENGINEERING

VOLUME 4

J. M. COULSON AND J. F. RICHARDSON

**Solutions to the Problems in Chemical Engineering
Volume 1**

By

J. R. BACKHURST AND J. H. HARKER

University of Newcastle upon Tyne

PERGAMON PRESS

OXFORD · NEW YORK · TORONTO · SYDNEY
PARIS · FRANKFURT

U.K.	Pergamon Press Ltd., Headington Hill Hall, Oxford OX3 0BW, England
U.S.A.	Pergamon Press Inc., Maxwell House, Fairview Park, Elmsford, New York 10523, U.S.A.
CANADA	Pergamon of Canada Ltd., 75 The East Mall, Toronto, Ontario, Canada
AUSTRALIA	Pergamon Press (Aust.) Pty. Ltd., 19a Boundary Street, Rushcutters Bay, N.S.W. 2011, Australia
FRANCE	Pergamon Press SARL, 24 rue des Ecoles, 75240 Paris, Cedex 05, France
WEST GERMANY	Pergamon Press GmbH, 6242 Kronberg-Taunus, Pferdstrasse 1, West Germany

Copyright © 1977 J. R. Backhurst and J. H. Harker

All Rights Reserved. No part of this publication may be reproduced, stored in a retrieval system or transmitted in any form or by any means: electronic, electrostatic, magnetic tape, mechanical, photocopying, recording or otherwise, without permission in writing from the publishers

First edition 1977

Library of Congress Cataloging in Publication Data

Coulson, John Metcalfe.
Chemical engineering.

Includes index.
CONTENTS: v. 1. Fluid flow, heat transfer, and mass transfer.
v. 4. Backhurst, J. R. and Harker, J. H. Solutions to the problems in Chemical engineering volume 1.
1. Chemical engineering. I. Richardson, John Francis joint author. II. Title.
TP155.C69 1976 660.2 75-42295
ISBN 0-08-020614-X (v. 1)
   0-08-020918-1 (Flexicover)
   0-08-020926-2 (Hardcover)

Printed in Great Britain by Biddles Ltd., Guildford, Surrey

CONTENTS

PREFACE

In the preface to the first edition of *Chemical Engineering*, Volume 1, Coulson and Richardson include the following paragraph:

"We have introduced into each chapter a number of worked examples which we believe are essential to a proper understanding of the methods of treatment given in the text. It is very desirable for a student to understand a worked example before tackling fresh practical problems himself. Chemical engineering problems require a numerical answer and it is essential to become familiar with the different techniques so that the answer is obtained by systematic methods rather than by intuition."

It is with these aims in mind that the present book, which in essence is a collection of solutions to the problems in the third edition of *Chemical Engineering*, Volume 1, has been prepared. The scope of the book is, of course, that of Volume 1, and the solutions are grouped in sections corresponding to the chapters in that text. The book has been written co-currently with the preparation of the new edition of Volume 1, and extensive reference has been made to the equations and sources of data in that volume at all stages. In this sense the present book is complementary to Volume 1. The working throughout is in SI units and the format is that of the third edition.

In common with countless students before us, we have battled with these problems for some two decades, and although our approach to the solutions has been refined over the years, we cannot claim to present the most elegant form of solution nor, indeed, always the most precise. Nevertheless, we hope that there is much to be learned from our efforts not only by the undergraduate but also by the professional engineer in industry.

This book could not have been written, of course, without the very real and longstanding contribution of Professors Coulson and Richardson to the profession. It is with considerable pleasure that we acknowledge our debt of gratitude, especially to Professor Coulson, who has guided our thoughts and encouraged our activities over so many years. We also acknowledge the help of our colleagues at the University of Newcastle upon Tyne and especially their forebearance during the preparation of this book.

Newcastle upon Tyne, 1976
J. R. Backhurst
J. H. Harker

UNITS AND DIMENSIONS

Problem 1.1

98% sulphuric acid of viscosity 0.025 N s/m^2 and density 1840 kg/m^3 is pumped at 685 cm^3/s through a 25 mm line. Calculate the value of the Reynolds number.

Solution

Cross-sectional area of line $= (\pi/4)0.025^2 = 0.00049$ m^2.
Mean velocity of acid, $u = 685 \times 10^{-6}/0.00049 = 1.398$ m/s.
Reynolds number, $Re = du\rho/\mu = 0.025 \times 1.398 \times 1840/0.025$
$\qquad\qquad\qquad\qquad = \underline{\underline{2572}}$

Problem 1.2

Compare the costs of electricity at 1p per kWh and town gas at 15p per therm.

Solution

Each cost is calculated in p/MJ.

$$1 \text{ kWh} = 1 \text{ kW} \times 1 \text{ h} = (1000 \text{ J/s})(3600 \text{ s})$$
$$= 3{,}600{,}000 \text{ J or } 3.6 \text{ MJ}$$
$$1 \text{ therm} = 105.5 \text{ MJ}$$

$\therefore\qquad$ cost of electricity $= 1$p/kWh or $1/3.6$ $\qquad = \underline{\underline{0.28\text{p/MJ}}}$
$\qquad\quad$ cost of gas $\quad = 15$p/105.5 MJ or $15/105.5 = \underline{\underline{0.14\text{p/MJ}}}$

Problem 1.3

A boiler plant raises 5.2 kg/s of steam at 1825 kN/m^2 pressure using coal of calorific value 27.2 MJ/kg. If the boiler efficiency is 75%, how much coal is consumed per day? If the steam is used to generate electricity, what is the power generation in kilowatts assuming a 20% conversion efficiency of the turbines and generators?

Solution

The total enthalpy of steam at 1825 kN/m^2 $= 2798$ kJ/kg (from steam tables).

$\therefore$ enthalpy of steam $= (5 \cdot 2 \times 2798) = 14{,}550 \text{ kW}$

Neglecting the enthalpy of the feed water, this enthalpy must be derived from the coal.

For an efficiency of 75%, heat provided by the coal $= (14550 \times 100/75) = 19{,}400 \text{ kW}$.

For a calorific value of $27{,}200 \text{ kJ/kg}$,

$$\text{rate of coal consumption} = (19{,}400/27{,}200) = 0 \cdot 713 \text{ kg/s}$$

or $(0 \cdot 713 \times 3600 \times 24/1000) = \underline{61 \cdot 6 \text{ Mg/day}}$

20% of the enthalpy in the steam is converted to power or

$$(14{,}550 \times 20/100) = 2910 \text{ kW or } 2 \cdot 91 \text{ MW}$$

$$\text{say } \underline{3 \text{ MW}}$$

Problem 1.4

The power required by an agitator in a tank is a function of the following four variables:

(a) diameter of impeller,
(b) number of rotations of impeller per unit time,
(c) viscosity of liquid,
(d) density of liquid.

From a dimensional analysis, obtain a relation between the power and the four variables.

The power consumption is found experimentally to be proportional to the square of the speed of rotation. By what factor would the power be expected to increase if the impeller diameter were doubled?

Solution

Let $P = \phi(DN\rho\mu)$ and hence a typical term of the function is $P = kD^a N^b \rho^c \mu^d$, where k is a constant.

The dimensions of each parameter in terms of $\mathbf{M}$, $\mathbf{L}$, and $\mathbf{T}$ are:

power, $P = \mathbf{ML^2/T^3}$ density, $\rho = \mathbf{M/L^3}$
diameter, $D = \mathbf{L}$ viscosity, $\mu = \mathbf{M/LT}$
speed of rotation, $N = \mathbf{T^{-1}}$

Equating dimensions:

$\mathbf{M}$: $1 = c + d$
$\mathbf{L}$: $2 = a - 3c - d$
$\mathbf{T}$: $-3 = -b - d$

Solving in terms of d:

$$a = (5 - 2d), \quad b = (3 - d), \quad c = (1 - d)$$

$$\therefore \qquad P = k\left(\frac{D^5}{D^{2d}}\frac{N^3}{N^d}\frac{\rho}{\rho^d}\mu^d\right)$$

or
$$P/D^5N^3\rho = k(D^2N\rho/\mu)^{-d}$$

or
$$N_P = kRe^m$$

Thus the power number is a function of the Reynolds number to the power m. In fact N_P is also a function of the Froude number, DN^2/g.

The above equation may be written as

$$P/D^5N^3\rho = k(D^2N\rho/\mu)^B$$

Experimentally
$$P \propto N^2$$

From the equation, $\qquad P \propto N^BN^3$, that is $B + 3 = 2$ and $B = -1$

Thus for the same fluid, that is the same viscosity and density,

$$(P_2/P_1)(D_1^5N_1^3/D_2^5N_2^3) = (D_1^2N_1/D_2^2N_2)^{-1}$$

or
$$(P_2/P_1) = (N_2^2D_2^3)/(N_1^2D_1^3)$$

In this case, $N_1 = N_2$ and $D_2 = 2D_1$.

$$\therefore \qquad (P_2/P_1) = 8D_1^3/D_1^3 = \underline{\underline{8}}$$

A similar solution may be obtained using the Recurring Set method as follows:

$$P = \phi(D, N, \rho, \mu)$$
$$f(P, D, N, \rho, \mu) = 0$$

Using **M**, **L** and **T** as fundamentals, there are five variables and three fundamentals and therefore by Buckingham's P_i theorem, there will be two dimensionless groups.

Choose D, N and ρ as recurring sets.

Dimensionally:

$$\left.\begin{array}{l}D \equiv \mathbf{L} \\ N \equiv \mathbf{T}^{-1} \\ \rho \equiv \mathbf{ML}^{-3}\end{array}\right] \qquad \text{Thus:} \left[\begin{array}{l}\mathbf{L} \equiv D \\ \mathbf{T} \equiv N^{-1} \\ \mathbf{M} \equiv \rho\mathbf{L}^3 \equiv \rho D^3\end{array}\right.$$

First group, π_1, is $P(\mathbf{ML^2T^{-3}})^{-1} \equiv P(\rho D^3D^2N^3)^{-1} \equiv \dfrac{P}{\rho D^5N^3}$

Second group, π_2, is $\mu(\mathbf{ML^{-1}T^{-1}})^{-1} \equiv \mu(\rho D^3D^{-1}N)^{-1} \equiv \dfrac{\mu}{\rho D^2N}$

Thus
$$f\left(\frac{P}{\rho D^5N^3}, \frac{\mu}{\rho D^2N}\right) = 0$$

Although there is little to be gained by using this method for simple problems, there certainly is when a large number of groups are involved.

Problem 1.5

It is found experimentally that the terminal settling velocity u_0 of a spherical particle in a fluid is a function of the following quantities:

  particle diameter, d,
  buoyant weight of particle (weight of particle − weight of displaced fluid), W,
  fluid density, ρ,
  fluid viscosity, μ.

Obtain a relationship for u_0 using dimensional analysis.

  Stokes established from theoretical considerations that for small particles which settle at very low velocities, the settling velocity is independent of the density of the fluid except in so far as this affects the buoyancy. Show that the settling velocity *must* then be inversely proportional to the viscosity of the fluid.

Solution

$$u_0 = k d^a W^b \rho^c \mu^d$$

Working in dimensions of **M**, **L** and **T**,

$$(\mathbf{L/T}) = k(\mathbf{L}^a(\mathbf{ML/T^2})^b(\mathbf{M/L^3})^c(\mathbf{M/LT})^d)$$

Equating dimensions:

  M: $0 = b + c + d$
  L: $1 = a + b - 3c - d$
  T: $-1 = -2b - d$

Solving in terms of B:

$$a = -1, \quad c = (b - 1), \quad \text{and } d = (1 - 2b)$$

∴ $u_0 = k(1/d)(W^b)(\rho^b/\rho)(\mu/\mu^{2b})$ where k is a constant,

or $\underline{u_0 = k(\mu/d\rho)(W\rho/\mu^2)^b}$

Rearranging:

$$(du_0\rho/\mu) = k(W\rho/\mu^2)^b$$

That is $(W\rho/\mu^2)$ is a function of a form of Reynolds number.
  For u_0 to be independent of ρ, b must equal unity and

$$\underline{u_0 = kW/d\mu}$$

or, for constant diameter and hence buoyant weight, the settling velocity is inversely proportional to the fluid viscosity.

ENERGY AND MOMENTUM RELATIONSHIPS

Problem 2.1

Calculate the ideal available energy produced by the discharge to atmosphere through a nozzle of air stored in a cylinder of capacity $0 \cdot 1 \, m^3$ at a pressure of $5 \, MN/m^2$. The initial temperature of the air is 290 K and the ratio of the specific heats is $1 \cdot 4$.

Solution

From equation 2.1, $dU = \delta q - \delta W$.
For an adiabatic process, $\delta q = 0$ and $dU = -\delta W$.
From equation 2.25, $dU = C_v \, dT = -\delta W$ for an isentropic process.

As $\gamma = C_p / C_v$ and $C_p = C_v + \mathbf{R}$ (from equation 2.27),

$$C_v = \mathbf{R}/(\gamma - 1)$$

$$W = -C_v \Delta T = -\mathbf{R}\Delta T/(\gamma - 1) = (\mathbf{R}T_1 - \mathbf{R}T_2)/(\gamma - 1)$$

Now
$$\mathbf{R}T_1 = P_1 v_1 \text{ and } \mathbf{R}T_2 = P_2 v_2$$

∴
$$W = (P_1 v_1 - P_2 v_2)/(\gamma - 1)$$

Now
$$P_1 v_1{}^\gamma = P_2 v_2{}^\gamma$$

and substituting for v_2 gives

$$W = [(P_1 v_1)/(\gamma - 1)](1 - (P_2/P_1)^{(\gamma - 1)/\gamma})$$

and
$$\Delta U = -W = [(P_1 v_1)/(\gamma - 1)][(P_2/P_1)^{(\gamma - 1)/\gamma} - 1]$$

In this problem,

$P_1 = 5 \, MN/m^2$, $P_2 = 0 \cdot 1013 \, MN/m^2$, $T_1 = 290 \, K$, and $\gamma = 1 \cdot 4$.
The specific volume, $v_1 = (22 \cdot 4/29)(290/273)(0 \cdot 1013/5) = 0 \cdot 0166 \, m^3/kg$.

∴
$$-W = [(5 \times 10^6 \times 0 \cdot 0166)/0 \cdot 4][(0 \cdot 1013/5)^{0 \cdot 4/1 \cdot 4} - 1]$$
$$= -0 \cdot 139 \times 10^6 \, J/kg$$

Mass of gas $= 0 \cdot 1/0 \cdot 0166 = 6 \cdot 02 \, kg$

∴
$$\Delta U = -0 \cdot 139 \times 10^6 \times 6 \cdot 20 = -0 \cdot 84 \times 10^6 \, J$$

or
$$\underline{\underline{-840 \, kJ}}$$

Problem 2.2

Obtain expressions for the variation of:

(a) internal energy with change of volume,
(b) internal energy with change of pressure, and
(c) enthalpy with change of pressure,

all at constant temperature, for a gas whose equation of state is given by van der Waal's law.

Solution

van der Waal's equation is given as equation 2.32 which for 1 kmol may be written as:

$$[P + (a/V^2)](V - b) = \mathbf{R}T$$

or

$$PV = \mathbf{R}T + bP - (a/V) + (ab/V^2)$$

or

$$P = [\mathbf{R}T/(V - b)] - (a/V^2)$$

(a) Equation 2.21 relates internal energy, pressure and temperature by:

$$\left(\frac{\partial U}{\partial V}\right)_T = T\left(\frac{\partial P}{\partial T}\right)_V - P$$

From van der Waal's equation,

$$\left(\frac{\partial P}{\partial T}\right)_V = \frac{\mathbf{R}}{(V - b)} \quad \text{and} \quad T\left(\frac{\partial P}{\partial T}\right)_V = \mathbf{R}T/(V - b)$$

Hence

$$\left(\frac{\partial U}{\partial V}\right)_T = \frac{\mathbf{R}T}{(V - b)} - P = \frac{a}{V^2}$$

NB—For an ideal gas, $b = 0$ and $(\partial U/\partial V)_T = (\mathbf{R}T/V) - P = 0$.

(b) From equation 2.23,

$$\left(\frac{\partial U}{\partial P}\right)_T = \left(\frac{\partial U}{\partial V}\right)_T\left(\frac{\partial V}{\partial P}\right)_T$$

$$(\partial V/\partial P) = 1/(\partial P/\partial V)$$

$$\frac{\partial P}{\partial V} = \frac{-\mathbf{R}T}{(V - b)^2} + \frac{2a}{V^3} = \frac{2a(V - b)^2 - \mathbf{R}TV^3}{V^3(V - b)^2}$$

$$\left(\frac{\partial U}{\partial V}\right)_T = \frac{\mathbf{R}T}{(V - b)} - P$$

∴

$$\left(\frac{\partial U}{\partial P}\right)_T = \left(\frac{\mathbf{R}T}{(V - b)} - P\right)\left(\frac{V^3(V - b)^2}{2a(V - b)^2 - \mathbf{R}TV^3}\right)$$

$$= \frac{[\mathbf{R}T - P(V - b)][V^3(V - b)^2]}{[2a(V - b)^2 - \mathbf{R}TV^3]}$$

NB—For an ideal gas, $a = b = 0$ and $(\partial U/\partial P)_T = 0$.

(c) From Volume 1, page 19;

$$dH = \left(\frac{\partial H}{\partial T}\right)_P dT + \left(\frac{\partial H}{\partial P}\right)_T dP$$

$$= C_p\, dT + \left(\frac{\partial U}{\partial P}\right)_T dP + \left(\frac{\partial (PV)}{\partial P}\right) dP$$

For a constant temperature process, $C_p\, dT = 0$ and

$$\frac{dH}{dP} = \left(\frac{\partial U}{\partial P}\right)_T + \left(\frac{\partial (PV)}{\partial P}\right)$$

$$\frac{\partial (PV)}{\partial T} = \frac{\partial}{\partial T}[RT + bP - (a/V) + (ab/V^2)]$$

$$= b$$

$$\therefore \qquad \frac{dH}{dP} = \frac{[RT - P(V - b)][V^3(V - b)^2]}{2a(V - b)^2 - RTV^3} - b$$

Problem 2.3

Calculate the energy stored in 1000 cm³ of gas at 80 MN/m² at 290 K using a datum of STP.

Solution

The key to the solution of this problem lies in the fact that the operation involved

is an irreversible expansion. As in problem 2.1, and taking C_v as constant between T_1 and T_2,

$$\Delta U = -W = nC_v(T_2 - T_1)$$

where n is the number of kmol of gas and T_2 and T_1 are the final and initial temperatures.

For a constant pressure process, the work done, assuming the ideal gas laws apply, is given by:

$$W = P_2(V_2 - V_1) = P_2\left(\frac{nRT_2}{P_2} - \frac{nRT_1}{P_1}\right)$$

Equating these expressions for W gives:

$$-C_v(T_2 - T_1) = P_2\left(\frac{RT_2}{P_2} - \frac{RT_1}{P_1}\right)$$

In this example,
$P_1 = 80000 \text{ kN/m}^2$, $P_2 = 101 \cdot 3 \text{ kN/m}^2$
$V_1 - 10^{-3} \text{ m}^3$, $R = 8\,314 \text{ kJ/kmol K}$, and $T_1 = 290 \text{ K}$
Hence

$$-C_v(T_2 - 290) = 101 \cdot 3 R[(T_2/101 \cdot 3) - (290/80000)]$$

If $\qquad C_v = 1 \cdot 5 R$ (that is $\gamma = 1 \cdot 66$), then $T_2 = 174 \cdot 15 \text{ K}$.

Now $PV = n\mathbf{R}T$ and $n = 80000 \times 10^{-3}/8 \cdot 314 \times 290 = 0 \cdot 033$ kmol

$$\Delta U = -W = C_v n (T_2 - T_1)$$
$$= 1 \cdot 5 \times 8 \cdot 314 \times 0 \cdot 033 (174 \cdot 15 - 290)$$
$$= -47 \cdot 7 \text{ kJ}$$

Problem 2.4

Compressed gas is distributed from a works in cylinders which are filled to a pressure P by connecting them to a large reservoir of gas which remains at a steady pressure P and temperature T. If the small cylinders are initially at a temperature T and pressure P_0, what is the final temperature of the gas in the cylinders if heat losses can be neglected and if the compression can be regarded as reversible? Assume that the ideal gas laws are applicable.

Solution

From equation 2.1, $\mathrm{d}U = \delta q - \delta W$.
For an adiabatic operation, $q = 0$ and $\delta q = 0$ and $\delta W = P \, \mathrm{d}v$ or $\mathrm{d}U = -P \, \mathrm{d}v$.
The change in internal energy for any process involving an ideal gas is given by equation 2.25:

$$C_v \, \mathrm{d}T = -P \, \mathrm{d}v = \mathrm{d}U$$

Now $P = \mathbf{R}T/v$ and hence $\mathrm{d}T/T = (-\mathbf{R}/C_v)(\mathrm{d}v/v)$

By definition, $\gamma = C_p/C_v$ and $C_p = C_v + \mathbf{R}$ from equation 2.27

$\therefore$ $\mathbf{R}/C_v = \gamma - 1$

and $\mathrm{d}T/T = -(\gamma - 1)(\mathrm{d}v/v)$

Integrating between conditions 1 and 2 gives:

$$\ln (T_2/T_1) = -(\gamma - 1) \ln (v_2/v_1)$$

or

$$(T_2/T_1) = (v_2/v_1)^{\gamma - 1}$$

Now $P_1 v_1/T_1 = P_2 v_2/T_2$ and hence $v_1/v_2 = (P_2/P_1)(T_1/T_2)$

and $(T_2/T_1) = (P_2/P_1)^{(\gamma - 1)/\gamma}$

In the symbols of the problem, the final temperature, $\underline{T_2 = T(P/P_0)^{(\gamma - 1)/\gamma}}$

SECTION 3

FRICTION IN PIPES AND CHANNELS

Problem 3.1

1250 cm^3/s of water is to be pumped through a steel pipe, 25 mm diameter and 30 m long, to a tank 12 m higher than its reservoir. Calculate approximately the power required. What type of pump would you install for the purpose and what power motor (in kW) would you provide?

$$\text{Viscosity of water} = 1\cdot30 \text{ mN s/m}^2$$
$$\text{Density of water} = 1000 \text{ kg/m}^3$$

Solution

For a 25 mm bore pipe, cross-sectional area $= (\pi/4)(0\cdot025)^2 = 0\cdot00049 \text{ m}^2$.
Velocity, $u = (1250 \times 10^{-6} \text{ m}^3/\text{s})/0\cdot00049 \text{ m}^2$
$= 2\cdot54 \text{ m/s}$
$Re = \rho u d/\mu$
$= 1000 \times 2\cdot54 \times 0\cdot025/1\cdot3 \times 10^{-3}$
$= 48900$

From Table 3.1 the roughness e of a steel pipe will be taken as $0\cdot045$ mm. Hence $e/d = 0\cdot046/25 = 0\cdot0018$.

From Fig. 3.7, when $e/d = 0\cdot0018$ and $Re = 4\cdot89 \times 10^4$, $R/\rho u^2 = 0\cdot0032$.

The pressure drop may be calculated from the energy balance equation and equation 3.16. For turbulent flow of an incompressible fluid:

$$\Delta u^2/2 + g\Delta z + v(P_2 - P_1) + 4(R/\rho u^2)(l/d)u^2 = 0$$

The pressure drop is equal to $P_1 - P_2$, i.e.

$$P_1 - P_2 = \rho[\Delta u^2/2 + g\Delta z + 4(R/\rho u^2)(l/d)u^2]$$
$$= \rho\{[0\cdot5 + 4(R/\rho u^2)(l/d)]u^2 + g\Delta z\}$$

since the velocity in the tank is equal to zero.

Substituting gives $P_1 - P_2 = 1000\{[0\cdot5 + 4 \times 0\cdot0032(30/0\cdot025)]2\cdot54^2 + 9\cdot81 \times 12\}$
$= 219,500 \text{ N/m}^2$
$= \underline{\underline{219\cdot5 \text{ kN/m}^2}}$

9

The power is obtained from equation 6.55:

$$\text{power} = G(u^2/2 + g\Delta z + F)$$
$$= (\text{kg/s})(\text{m}^2/\text{s}^2) = (\text{m}^3/\text{s})(\text{N/m}^2)$$
$$= (1 \cdot 25 \times 10^{-3})(2 \cdot 195 \times 10^5)$$
$$= 275 \text{ W}$$

If a pump efficiency of 60% is assumed, the pump motor should be rated at $275/0 \cdot 6 = \underline{\underline{458 \text{ W}}}$. A single stage centrifugal would be suitable for this duty.

Problem 3.2

Calculate the pressure drop in, and the power required to operate, a condenser consisting of 400 tubes, 4·5 m long and 10 mm internal diameter. The coefficient of contraction at the entrance of the tubes is 0·6, and 0·04 m³/s of water is to be pumped through the condenser.

Solution

The flow of water through each tube $= 0 \cdot 04/400 = 0 \cdot 0001 \text{ m}^3/\text{s}$.
Cross-sectional area of each tube $= (\pi/4)(0 \cdot 01)^2 = 0 \cdot 000785 \text{ m}^2$.
Water velocity $= 0 \cdot 0001/0 \cdot 000785 = 1 \cdot 273 \text{ m/s}$.
Equation 3.71 gives the entry pressure drop as:

$$\Delta P_f = -\frac{\rho u^2}{2}\left(\frac{1}{C_c} - 1\right)^2$$
$$= [1000 \times (1 \cdot 273)^2/2][(1/0 \cdot 6) - 1]^2$$
$$= 360 \text{ N/m}^2$$

$$\text{Reynolds number} = \rho u d/\mu = 1000 \times 1 \cdot 273 \times 0 \cdot 01/1 \cdot 0 \times 10^{-3}$$
$$= 1 \cdot 273 \times 10^4$$

If e is taken as 0·046 mm from Table 3.1, then $e/d = 0 \cdot 0046$ and from Fig. 3.7,
$$R/\rho u^2 = 0 \cdot 0043.$$
Equation 3.15 gives the pressure drop due to friction as;

$$\Delta P_f = 4(R/\rho u^2)(l/d)(\rho u^2)$$
$$= 4 \times 0 \cdot 0043(4 \cdot 5/0 \cdot 01)(1000 \times 1 \cdot 273^2)$$
$$= 12,540 \text{ N/m}^2$$

$$\text{Total pressure drop} = 12,540 + 360 = 12,900 \text{ N/m}^2$$
$$\equiv \underline{\underline{12 \cdot 9 \text{ kN/m}^2}}$$

Problem 3.3

75% sulphuric acid, of density 1650 kg/m^3 and viscosity $8 \cdot 6 \text{ mN s/m}^2$, is to be pumped for $0 \cdot 8$ km along a 50 mm internal diameter pipe at the rate of $3 \cdot 0$ kg/s, and then raised vertically 15 m by the pump. If the pump is electrically driven and has an efficiency of 50%, what power will be required? What type of pump would you use and of what material would you construct the pump and pipe?

Solution

Cross-sectional area of pipe $= (\pi/4)(0 \cdot 05)^2 = 0 \cdot 00196 \text{ m}^2$.
Velocity, $u = 3 \cdot 0/(1650 \times 0 \cdot 00196) = 0 \cdot 93$ m/s.
Reynolds number $= \rho u d / \mu = 1650 \times 0 \cdot 93 \times 0 \cdot 05/8 \cdot 6 \times 10^{-3}$
$$= 8900$$

If e is taken as $0 \cdot 046$ mm from Table 3.1, $e/d = 0 \cdot 00092$.
From Fig. 3.7, $R/\rho u^2 = 0 \cdot 0040$.
From equation 3.17 the head loss due to friction is given by:

$$h_f = -\Delta P_f/\rho g = 8(R/\rho u^2)(l/d)(u^2/2g)$$
$$= 8 \times 0 \cdot 004(800/0 \cdot 05)(0 \cdot 93^2/(2 \times 9 \cdot 81))$$
$$= 22 \cdot 6 \text{ m}$$

Total head $= 22 \cdot 6 + 15 = 37 \cdot 6$ m.
From equation 6·55: power $=$ mass flowrate $\times$ head $\times g$

$$= 3 \cdot 0 \times 37 \cdot 6 \times 9 \cdot 81$$
$$= 1105 \text{ W}$$

If the pump is 50% efficient, the power required $= 2210 \text{ W} = \underline{\underline{2 \cdot 2 \text{ kW}}}$.

For this duty a PTFE lined pump and lead piping would suffice.

Problem 3.4

60% sulphuric acid is to be pumped at the rate of 4000 cm^3/s through a lead pipe 25 mm diameter and raised to a height of 25 m. The pipe is 30 m long and includes two right-angled bends. Calculate the theoretical power required.

The specific gravity of the acid is $1 \cdot 531$ and its kinematic viscosity is $0 \cdot 425 \text{ cm}^2$/s. The density of water may be taken as 1000 kg/m^3.

Solution

Cross-sectional area of pipe $= (\pi/4)(0 \cdot 025)^2 = 0 \cdot 00049 \text{ m}^2$.
Velocity, $u = 4000 \times 10^{-6}/0 \cdot 00049 = 8 \cdot 15$ m/s.
Reynolds number $= \rho u d / \mu = u d / (\mu/\rho)$
$$= 8 \cdot 15 \times 0 \cdot 025/0 \cdot 425 \times 10^{-4}$$
$$= 4794$$

If e is taken as 0.05 mm from Table 3.1, $e/d = 0.002$ and from Fig. 3.7, $R/\rho u^2 = 0.0047$.

From equation 3.17, the head loss due to friction is given by:

$$h_f = 8(R/\rho u^2)(l/d)(u^2/2g)$$
$$= 8 \times 0.0047(30/0.025)(8.15^2/(2 \times 9.81))$$
$$= 152.8 \text{ m}$$

and $\Delta z = 25.0$ m

From Table 3.2, 0.8 velocity heads are lost through each $90°$ bend so that the loss through two bends $= 1.6$ velocity heads $= 1.6 \times 8.15^2/(2 \times 9.81) = 5.4$ m.
Total head loss $= 152.8 + 25 + 5.4 = 183.2$ m.
Mass flowrate $= 4000 \times 10^{-6} \times 1.531 \times 1000 = 6.12$ kg/s.
From equation 6.55, the theoretical power requirement $= 6.12 \times 183.2 \times 9.81$
$$= 11,000 \text{ W}$$
$$= \underline{\underline{11.0 \text{ kW}}}$$

Problem 3.5

1.3 kg/s of 98% sulphuric acid is to be pumped through a 25 mm diameter pipe, 30 m long, to a tank 12 m higher than its reservoir. Calculate the power required and indicate the type of pump and material of construction of the line that you would choose.

Viscosity of acid $= 0.025$ N s/m^2

Specific gravity $= 1.84$

Solution

Cross-sectional area of pipe $= (\pi/4)(0.0025)^2 = 0.00049$ m^2.
Volumetric flowrate $= 1.3/(1.84 \times 1000) = 0.00071$ m^3/s.
Velocity in the pipe, $u = 0.00071/0.00049 = 1.45$ m/s.
Reynolds number $= \rho u d/\mu$
$$= (1.84 \times 1000) \times 1.45 \times 0.025/0.025$$
$$= 2670$$

This value of the Reynolds number lies within the critical zone. If the flow were laminar, the value of $R/\rho u^2$ from Fig. 3.7 would be 0.003. If the flow were turbulent, the value of $R/\rho u^2$ would be considerably higher, and this higher value should be used in subsequent calculation to provide a margin of safety. If the roughness is taken to be 0.05 mm, $e/d = 0.05/25 = 0.002$ and $R/\rho u^2 = 0.0057$.

The head loss due to friction is then given by equation 3.17 as:

$$h_f = 8(R/\rho u^2)(l/d)(u^2/2g)$$
$$= 8 \times 0.0057(30/0.025)(1.45^2/(2 \times 9.81))$$
$$= 5.87 \text{ m}$$

In this example, $\Delta z = 12$ m, so that the total head $= 17.87$ m.
The theoretical power requirement is found from equation 6.55 as:

$$\text{power} = 17.87 \times 1.3 \times 9.81 = 227 \text{ W}$$

If the pump is 50% efficient, the actual power $= \underline{454 \text{ W}}$.

A PTFE lined centrifugal pump and lead or high silicon iron pipe would be suitable
for this duty.

Problem 3.6

Calculate the hydraulic mean diameter of the annular space between a 40 mm and
a 50 mm tube.

Solution

The hydraulic mean diameter d_m is defined as four times the cross-sectional area
divided by the wetted perimeter. Equation 3.61 gives the value d_m for an annulus of
outer radius r and inner radius r_i as:

$$d_m = 4\pi(r^2 - r_i^2)/2\pi(r + r_i)$$
$$= 2(r - r_i)$$

Hence if $\qquad\qquad r = 25$ mm $\quad$ and $\quad r_i = 20$ mm

$$d_m = 2(25 - 20) = \underline{10 \text{ mm}}$$

Problem 3.7

0.015 m³/s of acetic acid is pumped through a 75 mm diameter horizontal pipc 70 m
long. What is the pressure drop in the pipe?

Viscosity of acid	$= 2.5$ mN s/m²
Density of acid	$= 1060$ kg/m³
Roughness of pipe surface	$= 6 \times 10^{-5}$ m

Solution

Cross-sectional area of pipe $= (\pi/4)(0.075)^2 = 0.0044$ m².
Velocity of acid in the pipe, $u = 0.015/0.0044 = 3.4$ m/s.
Reynolds number $= \rho u d/\mu = 1060 \times 3.4 \times 0.07/2.5 \times 10^{-3}$
$$= 1.08 \times 10^5$$
Pipe roughness $e = 6 \times 10^{-5}$ m and $e/d = 6 \times 10^{-5}/0.075$
$$= 0.0008$$

The pressure drop is calculated from equation 3.15 as:

$$\Delta P_f = 4(R/\rho u^2)(l/d)(\rho u^2)$$

From Fig. 3.7, when $Re = 1\cdot08 \times 10^5$ and $e/d = 0\cdot0008$, $R/\rho u^2 = 0\cdot0025$. Substitution gives

$$\Delta P_f = 4 \times 0\cdot0025(70/0\cdot075)(1060 \times 3\cdot4^2)$$
$$= 11,4370 \text{ N/m}^2$$
$$\equiv \underline{\underline{114\cdot4 \text{ kN/m}^2}}$$

Problem 3.8

A cylindrical tank, 5 m in diameter, discharges through a mild steel pipe 90 m long and 230 mm diameter connected to the base of the tank. Find the time taken for the water level in the tank to drop from 3 m to 1 m above the bottom.

Take the viscosity of water as 1 mN s/m².

Solution

At any time let the depth of water in the tank be h and let levels 1 and 2 be the liquid level in the tank and the pipe outlet respectively.

The energy balance equation states:

$$\Delta(u^2/2) + g\Delta z + v(P_2 - P_1) + F = 0$$

In this example, $P_1 = P_2 = $ atmospheric pressure and $v(P_2 - P_1) = 0$. Also $u_1/u_2 = (0\cdot23/5)^2 = 0\cdot0021$ so that u_1 may be neglected. The energy balance equation then becomes:

$$u^2/2 - hg + 4(R/\rho u^2)(l/d)u^2 = 0$$

The last term is obtained from equation 3.16 and $\Delta z = -h$. Substituting the known data gives:

$$u^2/2 - 9\cdot81h + 4(R/\rho u^2)(90/0\cdot23)u^2 = 0$$

or

$$u^2 - 19\cdot62h + 3130(R/\rho u^2)u^2 = 0$$

from which

$$u = 4\cdot43\sqrt{h}/\sqrt{[1 + 3130(R/\rho u^2)]}$$

In falling from a height h to $h - dh$, the quantity of water discharged = $(\pi/4)5^2(-dh) = 19\cdot63dh$ m³.

Volumetric flow rate = $(\pi/4)(0\cdot23)^2 u = 0\cdot0415u = 0\cdot184\sqrt{h}/\sqrt{[1 + 3130(R/\rho u^2)]}$.

Hence the time taken for the level to fall from h to $h - dh$ is

$$= \frac{-19\cdot63}{0\cdot184} \frac{dh}{\sqrt{h}} \sqrt{[1 + 3130(R/\rho u^2)]}$$

$$= -106\cdot7h^{-0\cdot5}\sqrt{[1 + 3130(R/\rho u^2)]}dh$$

The time taken for the level to fall from 3 m to 1 m is given by

$$t = -106 \cdot 7\sqrt{[1 + 3130(R/\rho u^2)]} \int_{3}^{1} h^{-0.5} dh$$

Now $R/\rho u^2$ depends upon the Reynolds number which will fall as the level in the tank falls. It also depends upon the roughness of the pipe e which is not specified in this example.

The pressure drop along the pipe $= h\rho g = 4Rl/d \ \text{N/m}^2$ and $R = h\rho g d/4l$.

From equation 3.20:

$$(R/\rho u^2)Re^2 = Rd^2\rho/\mu^2 = h\rho^2 g d^3/4l\mu^2$$

$$= h \times 1000^2 \times 9\cdot 81 \times 0\cdot 23^3/4 \times 90 \times 10^{-6}$$

$$= 3\cdot 315 \times 10^8 h$$

Thus as h varies from 3 m to 1 m, $(R/\rho u^2)(Re)^2$ varies from $9\cdot 95 \times 10^8$ to $3\cdot 315 \times 10^8$.

If $R/\rho u^2$ is taken as $0\cdot 002$, Re will vary from $7\cdot 05 \times 10^5$ to $4\cdot 07 \times 10^5$. From Fig. 3.7 this corresponds to a range of e/d of between $0\cdot 004$ and $0\cdot 005$ or a roughness of between $0\cdot 92$ and $1\cdot 15$ mm, which is too high for a commercial pipe.

If e is taken to be $0\cdot 05$ mm, $e/d = 0\cdot 0002$, and for Reynolds numbers near 10^6, $R/\rho u^2 = 0\cdot 00175$. Substituting $R/\rho u^2 = 0\cdot 00175$ and integrating gives a time of $99\cdot 4$ s for the level to fall from 3 m to 1 m. If $R/\rho u^2 = 0\cdot 00175$, Re varies from $7\cdot 5 \times 10^5$ to $4\cdot 35 \times 10^5$, and from Fig. 3.7, $e/d = 0\cdot 00015$, which is near enough to the assumed value.

Thus the time for the level to fall is approximately <u>100 s</u>.

Problem 3.9

Two storage tanks A and B containing a petroleum product discharge through pipes each $0\cdot 3$ m in diameter and $1\cdot 5$ km long to a junction at D. From D the product is carried by a $0\cdot 5$ m diameter pipe to a third storage tank, C, $0\cdot 8$ km away. The surface of the liquid in A is initially 10 m above that in C and the liquid level in B is 7 m higher than that in A. Calculate the initial rate of discharge of the liquid if the pipes are of mild steel. Take the density of the petroleum product as 870 kg/m³ and the viscosity as $0\cdot 7$ mN s/m².

Solution

The pipe network is shown in Fig. 3a and a datum corresponding to the level in tank C will be taken. Kinetic energy and entry losses may be neglected as all the pipes are long.

Let u_1, u_2, and u_3 be the velocities in AD, BD, and DC respectively, and assume

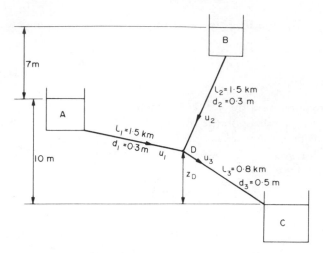

initially that $R/\rho u^2 (= \phi)$ is the same in each pipe. Let the pressure at D be P_D N/m² and z_D m the height of D above the datum level.

The energy balance equation (2.67) is now applied between D and the liquid level in each tank.

For AD $\qquad (z_D - 10)g + vP_D + 4\phi(1500/0\cdot3)u_1^2 = 0 \qquad$ (1)

For BD $\qquad (z_D - 17)g + vP_D + 4\phi(1500/0\cdot3)u_2^2 = 0 \qquad$ (2)

For DC $\qquad - z_Dg - vP_D + 4\phi(800/0\cdot5)u_3^2 = 0 \qquad$ (3)

Substracting (2) from (1) gives:

$$7g + 20,000\phi(u_1^2 - u_2^2) = 0 \qquad (4)$$

Adding (2) and (3) gives:

$$-17g + 20,000\phi(u_2^2 + 0\cdot32u_3^2) = 0 \qquad (5)$$

If $R/\rho u^2$ is taken as $0\cdot002$, equations (4) and (5) become:

$$68\cdot67 + 40(u_1^2 - u_2^2) = 0 \qquad (6)$$
$$-166\cdot77 + 40(u_2^2 - 0\cdot32u_2^2) = 0 \qquad (7)$$

The flow in DC is the sum of the flows in AD and BD, i.e.

$$(\pi/4)0\cdot03^2(u_1 + u_2) = (\pi/4)0\cdot5^2u_3$$

and $\qquad\qquad\qquad\qquad u_1 + u_2 = 2\cdot78u_3 \qquad$ (8)

From equation (6), $\qquad\qquad\qquad u_1^2 = u_2^2 - 1\cdot72 \qquad$ (9)

Equations (7) and (9) may now be solved to give u_1:

$$-166\cdot77 + 40\{u_2^2 + 0\cdot32[0\cdot13(\sqrt{u_2^2 - 1\cdot72} + u_2)^2]\} = 0$$

and $\qquad\qquad\qquad\qquad u_2 = 1\cdot96$ m/s

From equation (9), $u_1^2 = 1 \cdot 96^2 - 1 \cdot 72 = 1 \cdot 46$ m/s

From equation (8), $u_3 = 0 \cdot 36(u_1 + u_2) = 1 \cdot 23$ m/s

The Reynolds number may now be calculated in each pipe:

AD $Re = 870 \times 1 \cdot 46 \times 0 \cdot 3/0 \cdot 7 \times 10^{-3} = 5 \cdot 44 \times 10^5$

BD $Re = 870 \times 1 \cdot 96 \times 0 \cdot 3/0 \cdot 7 \times 10^{-3} = 7 \cdot 31 \times 10^5$

DC $Re = 870 \times 1 \cdot 23 \times 0 \cdot 5/0 \cdot 7 \times 10^{-3} = 7 \cdot 64 \times 10^5$

If e is taken as $0 \cdot 046$ mm, $e/d = 0 \cdot 046/300$ or $0 \cdot 046/500$, say an average of $0 \cdot 0001$. From Fig. 3.7, for $e/d = 0 \cdot 0001$ and Re between $5 \cdot 4$ and $7 \cdot 6 \times 10^5$, $R/\rho u^2 = 0 \cdot 002$. Thus the initial assumption for $R/\rho u^2$ was satisfactory.

The rate of discharge $= (\pi/4)0 \cdot 5^2 \times 1 \cdot 23 \times 870$

$$= \underline{\underline{170 \cdot 8 \text{ kg/s}}}$$

Problem 3.10

Find the drop in pressure due to friction in a pipe 300 m long and 150 mm diameter when water is flowing at the rate of $0 \cdot 05$ m³/s. The pipe is of glazed porcelain.

Solution

For a glazed porcelain pipe, take $e = 0 \cdot 0015$ mm, $e/d = 0 \cdot 0015/150 = 0 \cdot 00001$.

Cross-sectional area of pipe $= (\pi/4)(0 \cdot 15)^2 = 0 \cdot 0176$ m².

Velocity of water in pipe, $u = 0 \cdot 05/0 \cdot 0176 = 2 \cdot 83$ m/s.

Reynolds number $= \rho u d/\mu = 1000 \times 2 \cdot 83 \times 0 \cdot 15/10^{-3}$
$$= 4 \cdot 25 \times 10^5$$

From Fig. 3.7, $R/\rho u^2 = 0 \cdot 0017$.

The pressure drop is given by equation 3.15 as:

$$\Delta P_f = 4(R/\rho u^2)(l/d)(\rho u^2)$$
$$= 4 \times 0 \cdot 0017(300/0 \cdot 15)(1000 \times 2 \cdot 83^2)$$
$$= 108,900 \text{ N/m}^2$$
$$\simeq \underline{\underline{1 \text{ MN/m}^2}}$$

Problem 3.11

Two tanks, the bottoms of which are at the same level, are connected with one another by a horizontal pipe 75 mm diameter and 300 m long. The pipe is bell-mouthed at each end so that losses on entry and exit are negligible. One tank is 7 m diameter and contains water to a depth of 7 m. The other tank is 5 m diameter and contains water to a depth of 3 m.

If the tanks are connected to each other by means of the pipe, how long will it take before the water level in the larger tank has fallen to 6 m? Assume the pipe to be an old mild steel pipe.

Solution

The system is shown in Fig. 3b. If at any time t the depth of water in the larger tank is h and the depth in the smaller tank is H, a relationship between h and H may be found.

Area of larger tank $= (\pi/4)7^2 = 38\cdot48$ m^2.
Area of smaller tank $= (\pi/4)5^2 = 19\cdot63$ m^2.

When the level in the large tank falls to h, the volume discharged $=$ $(7-h)\times 38\cdot48$ m^3. The level in the small tank will rise by a height x where x is given by:

$$x = 34\cdot48(7-h)/19\cdot63$$
$$= 13\cdot72 - 1\cdot95h$$
$$H = x + 3 = 16\cdot72 - 1\cdot95h$$

The energy balance equation is:

$$\Delta u^2/2 + g\,\Delta z + v(P_1 - P_2) + F = 0$$

$\Delta u^2/2$ may be neglected, and $P_1 - P_2 =$ atmospheric pressure, so that:

$$g\,\Delta z + F = 0 = g\,\Delta z + 4(R/\rho u^2)(l/d)u^2$$
$$\Delta z = h - H$$
$$= 0\cdot95h - 16\cdot72$$

$\therefore$ $$(0\cdot95h - 16\cdot72)g + 4(R/\rho u^2)(l/d)u^2 = 0$$

and $$u = \sqrt{[(16\cdot72 - 0\cdot95h)g/4(R/\rho u^2)(l/d)]}$$

As the level falls from h to $h-dh$ in time dt, the volume discharged $=$ $-38\cdot48(dh)$ m^3.

Hence   time $dt = \dfrac{-38\cdot48dh}{(\pi/4)(0\cdot075)^2\sqrt{[(16\cdot72 - 0\cdot95h)g/4(R/\rho u^2)(l/d)]}}$

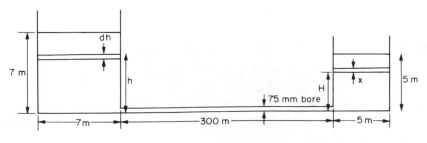

FIG. 3b

i.e.
$$dt = \frac{-2780dh\sqrt{[4(R/\rho a^2)(l/d)]}}{\sqrt{(16\cdot72 - 0\cdot95h)}}$$

If $R/\rho u^2$ is taken to be $0\cdot002$,

$$\int dt = 15740 \int_{7}^{6} \frac{dh}{\sqrt{(16\cdot72 - 0\cdot95h)}}$$

and
$$t = 4750\ s$$

The average volumetric flowrate $= 38\cdot48\ (7 - 6)/4750$
$$= 0\cdot0081\ m^3/s$$
Cross-sectional area of pipe $= 0\cdot00442\ m^2$.
Average velocity in the pipe $= 0\cdot0081/0\cdot00442 = 1\cdot83\ m/s$.
Reynolds number $= 1000 \times 1\cdot83 \times 0\cdot075/10^{-3} = 1\cdot36 \times 10^5$.
From Fig. 3.7, if $e = 0\cdot05\ mm$, $e/d = 0\cdot00067$, $R/\rho u^2 = 0\cdot0023$, which is near
enough to the assumed value of $0\cdot002$.
Thus the time for the level to fall $= \underline{4750\ s}$.

Problem 3.12

Two immiscible fluids A and B, of viscosities μ_A and μ_B, flow under streamline
conditions between two horizontal parallel planes of width b, situated a distance $2a$
apart (where a is much less than b), as two distinct parallel layers one above the
other, each of depth a as shown in Fig. 3c.
Show that the volumetric rate of flow of A is

$$\frac{\Delta P a^3 b}{12\mu_A l}\left(\frac{7\mu_A + \mu_B}{\mu_A + \mu_B}\right)$$

where ΔP is the pressure drop over a length l in the direction of flow.

Solution

A force balance over an element distance y from the axis gives:

$$\Delta P\pi y^2 = -\mu\,(du/dy)2\pi yl \text{ from which}$$

for fluid A: $\mu_A(d^2u_{xA}/dy^2) = -\Delta P/l$
for fluid B: $\mu_B(d^2u_{xB}/dy^2) = -\Delta P/l$

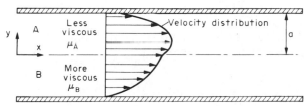

FIG. 3c

Integrating gives:
$$u_{xB} = \frac{-\Delta P}{2\mu_B l} x^2 + C_{1B} x + C_{2B}$$

and
$$u_{xA} = \frac{-\Delta P}{2\mu_A l} x^2 + C_{1A} x + C_{2A}$$

where C_{1A}, C_{1B}, C_{2A}, and C_{2B} are constants of integration. These constants may be determined by considering the boundary conditions:

(a) Fluid A, $\qquad\qquad\qquad\qquad u_{xA} = 0 \quad$ at $\quad y = -a$

(b) Fluid B, $\qquad\qquad\qquad\qquad u_{xB} = 0 \quad$ at $\quad y = +a$

(c) Both A and B have equal velocities at the axis, i.e.

$$u_{xA} = u_{xB} \quad \text{at} \quad y = 0$$

(d) The transport of x-momentum in the y-direction must be the same at $y = 0$ whether it is calculated from the velocity profile in region A or from the velocity profile in region B. If R represents the shear stress then:

$$R_{yxA} = R_{yxB}$$

or
$$-\mu_A (du_A/dy) = -\mu_B (du_B/dy) \text{ at } x = 0$$

From (a), $\qquad\qquad\qquad 0 = \frac{-\Delta P}{2\mu_B l} a^2 - C_{1B} a + C_{2B}$

From (b), $\qquad\qquad\qquad 0 = \frac{-\Delta P}{2\mu_A l} a^2 + C_{1A} a + C_{2A}$

From (c), $\qquad\qquad\qquad\qquad C_{2B} = C_{2A}$

From (d), $\qquad\qquad\qquad\qquad -\mu_B C_{1B} = -\mu_A C_{1A}$

Hence, $\qquad\qquad C_{1B} = \frac{\Delta P a}{2\mu_B l} \left(\frac{\mu_B - \mu_A}{\mu_B + \mu_A} \right) = C_{1A} \frac{\mu_A}{\mu_B}$

and $\qquad\qquad C_{2B} = \frac{\Delta P a^2}{2\mu_B l} \left(\frac{2\mu_B}{\mu_B + \mu_A} \right) = C_{2A}$

Hence the velocity distributions may now be calculated as:

$$u_{xA} = \frac{\Delta P a^2}{2\mu_A l} \left[\left(\frac{2\mu_A}{\mu_A + \mu_B} \right) + \left(\frac{\mu_B - \mu_A}{\mu_B + \mu_A} \right) \left(\frac{y}{a} \right) - \left(\frac{y}{a} \right)^2 \right]$$

$$u_{xB} = \frac{\Delta P a^2}{2\mu_B l} \left[\left(\frac{2\mu_B}{\mu_B + \mu_A} \right) + \left(\frac{\mu_B - \mu_A}{\mu_B + \mu_A} \right) \left(\frac{y}{a} \right) - \left(\frac{y}{a} \right)^2 \right]$$

NB—If $\mu_A = \mu_B$, both velocity distributions are the same.

The average velocity of fluid A layer is:

$$u_{A \text{ av}} = \frac{1}{a} \int_0^a u_{xA} \, dy = \frac{\Delta P a^2}{12\mu_A l} \left(\frac{\mu_B + 7\mu_A}{\mu_B + \mu_A} \right)$$

Volumetric flowrate

$$= u_{A\,av} \times ab$$

$$= \frac{\Delta P a^3 b}{12\mu_A}\left(\frac{\mu_B + 7\mu_A}{\mu_B + \mu_A}\right)$$

Problem 3.13

A petroleum fraction is pumped 2 km from a distillation plant to storage tanks through a mild steel pipeline, 150 mm in diameter, at the rate of 0·04 m³/s. What is the pressure drop along the pipe and the power supplied to the pumping unit if it has an efficiency of 50%?

The pump impeller is eroded and the pressure at its delivery falls to one half. By how much is the flowrate reduced?

Specific gravity of the liquid = 0·705
Viscosity of the liquid   = 0·5 mN s/m²
Roughness of pipe surface   = 0·004 mm

Solution

Cross-sectional area of pipe = $(\pi/4)0\cdot15^2 = 0\cdot0177$ m².
Velocity in the pipe = 0·04/0·0177 = 2·26 m/s.
Reynolds number = $0\cdot705 \times 1000 \times 2\cdot26 \times 0\cdot15/0\cdot5 \times 10^{-3}$
$\qquad\qquad\qquad = 4\cdot78 \times 10^5$
$e = 0\cdot004$ mm, $e/d = 0\cdot004/150 = 0\cdot000027$.
From Fig. 3.7, $R/\rho u^2 = 0\cdot00165$.
The pressure drop is given by equation 3.15,

$$\Delta P_f = 4(R/\rho u^2)(l/d)(\rho u^2)$$
$$= 4 \times 0\cdot00165(2000/0\cdot15)(0\cdot705 \times 1000 \times 2\cdot26^2)$$
$$= 316,900 \text{ N/m}^2$$
$$\simeq 320 \text{ kN/m}^2$$

The power required if the pump has an efficiency of 50% is:

power = (head × mass flowrate × g)/0·5
$\qquad$ = pressure drop (N/m²) × volumetric flowrate (m³/s)/0·5
$\qquad$ = 316,900 × 0·04/0·5
$\qquad$ = 25,350 W = __25·4 kW__

If, due to impeller erosion, the delivery pressure is halved, the new flowrate may be found from equation 3.20 and Fig. 3.8:

$$(R/\rho u^2)Re^2 = \Delta P_f d^3 \rho/4l\mu^2$$

The new pressure drop $= 316,900/2 = 158,450 \text{ N/m}^2$ and,

$$(R/\rho u^2)Re^2 = (158,450 \times 0\cdot15^3 \times 705)/(4 \times 2000 \times 0\cdot5^2 \times 10^{-6})$$
$$= 1\cdot885 \times 10^8$$

From Fig. 3.8, when $(R/\rho u^2)Re^2 = 1\cdot9 \times 10^8$ and $e/d = 0\cdot000027$, $Re = 3\cdot0 \times 10^5$,

i.e. $3\cdot0 \times 10^5 = 705 \times 0\cdot15 \times u/0\cdot5 \times 10^{-3}$

and $u = 1\cdot418 \text{ m/s}$

The volumetric flowrate is now $\underline{\underline{1\cdot418 \times 0\cdot0177 = 0\cdot025 \text{ m}^3/\text{s}}}$

Problem 3.14

Glycerol is pumped from storage tanks to rail cars through a single 50 mm diameter main 10 m long, which must be used for all grades of glycerol. After the line has been used for commercial material, how much pure glycerol must be pumped before the issuing liquid contains not more than 1% of the commercial material? The flow in the pipeline is streamline and the two grades of glycerol have identical densities and viscosities.

Solution

A force balance over an element distance r from the axis of a pipe whose radius is a, is given by:

$$\Delta P\pi r^2 = -\mu(du/dr)2\pi rl$$

where u is the velocity at distance r and l is the length of the pipe.

Hence $du = -(\Delta P/2\mu l)r \ dr$

$\therefore$ $u = -(\Delta P/4\mu l)r^2 + \text{constant}$

When $r = a$, $u = 0$ and constant $= (\Delta P/4\mu l)a^2$

$\therefore$ $u = (\Delta P/4\mu l)(a^2 - r^2)$

At a distance r from the axis, the time taken for the fluid to flow through a length l is given by $4\mu l^2/\Delta P(a^2 - r^2)$.

The volumetric rate of flow from $r = 0$ to $r = r$

$$= \int_0^r (\Delta P/4\mu l)(a^2 - r^2)2\pi r \ dr$$

$$= (\pi \Delta P/8\mu l)(2a^2r^2 - r^4)$$

Volumetric flowrate over the whole pipe $= \pi\Delta Pa^4/8\mu l$ and the mean velocity $= \Delta Pa^2/8\mu l$.

The required condition at the pipe outlet is

$$\frac{(\pi \Delta P/8\mu l)(2a^2 r^2 - r^4)}{\pi \Delta P a^4/8\mu l} = 0 \cdot 99$$

from which $r = 0 \cdot 95a$.

The time for fluid at this radius to flow through length l

$$= 4\mu l^2/\Delta P a^2 (1 - 0 \cdot 95^2) = 41\mu l^2/\Delta P a^2$$

Hence volume to be pumped $= (41\mu l^2/\Delta P a^2)(\pi \Delta P a^4/8\mu l)$
$$= 41\pi a^2/8 = 41\pi (0 \cdot 025)^2/8$$
$$= 0 \cdot 10 \text{ m}^3$$

Problem 3.15

A viscous fluid flows through a pipe with slightly porous walls so that there is a leakage of kP m³/m² s where P is the local pressure, measured above the discharge pressure, and k is a constant. After a length L, the liquid is discharged into a tank.

If the internal diameter of the pipe is D m and the volumetric rate of flow at the inlet is Q m³/s, show that the pressure drop in the pipe is given by:

$$P = \frac{Q}{\pi kD} a \tanh aL$$

where $\qquad a = (128 \, k\mu /D^3)^{0 \cdot 5}$

Assume a fully developed flow with $(R/\rho u^2) = 8Re^{-1}$.

Solution

The pressure drop over an incremental length dL is given by:

$$dP = 4(R/\rho u^2)(dL/D)\rho u^2$$

In this case, $\qquad R/\rho u^2 = 8/Re = 8\mu/Du\rho$

and $\qquad dP = 32\mu u \, dL/D^2$

or $\qquad dP/dL = 32\mu u/D^2 \qquad\qquad (1)$

The leakage from the element $= kP\pi D \, dL$ m³/s, and hence the change in velocity $du = kP\pi D \, dl/(\pi/4)D^2$

or $\qquad du/dL = 4kP/D \qquad\qquad (2)$

From (1) $\qquad d^2P/dL^2 = (32\mu/D^2)du/dL$

and substituting from (2),

$$d^2P/dL^2 = (128\mu k/D^3)P$$

or $\qquad d^2P/dL^2 = a^2 P$

where $\qquad a = (128\mu k/D^3)^{0 \cdot 5}$

The general solution to this equation is:

$$P = A_1 \cosh aL + A_2 \sinh aL \qquad (3)$$

When $L = 0$ and $P = 0$, substitution in (3) gives

$$0 = A_1$$

$$\therefore \qquad P = A_2 \sinh aL$$

$$dP/dL = A_2 a \cosh aL = 32\mu u/D^2 \quad \text{from (1)} \qquad (4)$$

when $L = L$, $u = Q/(\pi/4)D^2$ and $dP/dL = a^2 Q/\pi kD$,

and in (4), $\qquad\qquad\qquad a^2 Q/\pi kD = A_2 a \cosh aL$

$$\therefore \qquad\qquad\qquad A_2 = a(Q/\pi kD)\cosh aL$$

and in (3), $\qquad\qquad\qquad P = a(Q/\pi kD)\sinh aL/\cosh aL$

$$\underline{\underline{= (Q/\pi kD)a \tanh aL}}$$

Problem 3.16

A petroleum product of viscosity 0.5 mN s/m^2 and specific gravity 0.7 is pumped through a pipe of 0.15 m diameter to storage tanks situated 100 m away. The pressure drop along the pipe is 70 kN/m^2. The pipeline has to be repaired and it is necessary to pump the liquid by an alternative route consisting of 70 m of 200 mm pipe followed by 50 m of 100 mm pipe. If the existing pump is capable of developing a pressure of 300 kN/m^2, will it be suitable for use during the period required for the repairs? Take the roughness of the pipe surface as 0.005 mm.

Solution

This problem will be solved by using equation 3.20 and Fig. 3.8 to find the volumetric flowrate and then calculating the pressure drop through the alternative pipe system.

From equation 3.20, $\qquad (R/\rho u^2)Re^2 = \Delta P_f d^3 \rho/4l\mu^2$

$$= (70{,}000 \times 0.15^3 \times 700)/(4 \times 100 \times 0.5^2 \times 10^{-6})$$

$$= 1.65 \times 10^9$$

From Fig. 3.8, $Re = 8.8 \times 10^5 = 700 \times 0.15u/0.5 \times 10^{-3}$ and the velocity $u = 3.19$ m/s.

Cross-sectional area $= (\pi/4)0.15^2 = 0.0177$ m^2.

Volumetric flowrate $= 4.19 \times 0.0177 = 0.074$ m^3/s.

The velocity in the 0.2 m diameter pipe $= 0.074/(\pi/4)0.2^2$

$$= 2.36 \text{ m/s}$$

The velocity in the 0.1 m diameter pipe $= 9.44$ m/s.

Reynolds number in the 0.2 m pipe $= 700 \times 2.36 \times 0.2/0.5 \times 10^{-3}$

$$= 6.6 \times 10^5$$

Reynolds number in the 0·1 m pipe $= 700 \times 9\cdot44 \times 0\cdot1/0\cdot5 \times 10^{-3}$
$$= 1\cdot32 \times 10^6$$

e/d for the 0·2 m and the 0·1 m pipes $= 0\cdot00025$ and $0\cdot0005$ respectively.
From Fig. 3.7, $R/\rho u^2 = 0\cdot0018$ and $0\cdot002$ respectively, and from equation 3.15,

$$\Delta P_f = [4 \times 0\cdot0018(70/0\cdot2)(700 \times 2\cdot36^2)] + [4 \times 0\cdot002(50/0\cdot1)(700 \times 9\cdot42^2)]$$
$$= 258{,}300 \ \text{N/m}^2 = \underline{258\cdot3 \ \text{kN/m}^2}$$

Thus the existing pump is satisfactory for this duty.

Problem 3.17

Explain the phenomenon of hydraulic jump which occurs during the flow of a liquid in an open channel.

A liquid discharges from a tank into an open channel under a gate so that the liquid is initially travelling at a velocity of 1·5 m/s and a depth of 75 mm. Calculate, from first principles, the corresponding velocity and depth after the jump.

Solution

The topic of hydraulic jump is covered in Section 3.6.4 where it is shown from first principles that the depth of fluid in the channel D_2 is given by equation 3.98:

$$D_2 = 0\cdot5\{-D_1 + \sqrt{[D_1^2 + (8u_1^2 D_1/g)]}\}$$

where D_1 and u_1 are the depth and velocity before the jump.

In this example, $D_1 = 0\cdot075$ m and $u_1 = 1\cdot5$ m/s, so that:

$$D_2 = 0\cdot5\{-0\cdot075 + \sqrt{[0\cdot075^2 + (8 \times 1\cdot5^2 \times 0\cdot075/9\cdot81)]}\}$$
$$= \underline{0\cdot152 \ \text{m}}$$

If the channel has a uniform cross-sectional area,

$$u_1 D_1 = u_2 D_2$$

and $\qquad\qquad u_2 = u_1 D_1/D_2 = 1\cdot5 \times 0\cdot075/0\cdot152 = \underline{0\cdot74 \ \text{m/s}}$

Problem 3.18

What is a non-Newtonian fluid? Describe the principal types of behaviour exhibited by these fluids. The viscosity of a non-Newtonian fluid changes with the rate of shear according to the approximate relationship:

$$\mu = \mu_0 \frac{du}{dr}$$

where μ is the viscosity, and du/dr is the velocity gradient normal to the direction of motion.

Show that the volumetric rate of streamline flow through a horizontal tube of radius a is:

$$(2\pi/7)a^{3.5}\sqrt{-\Delta P/2l\mu_0}$$

where ΔP is the pressure drop over a length l of the tube.

Solution

Non-Newtonian fluids are discussed in Section 3.3 where a description of their behaviour under shear conditions is given.

A force balance over an element of radius r gives:

$$-\Delta P\pi r^2 = R2\pi rl$$

and
$$R = \mu_0(du/dr)^2$$

Hence
$$(du/dr)^2 = -\Delta P/2l\mu_0$$

$$du/dr = -\sqrt{-\Delta P/2l\mu_0}$$

Integrating gives
$$u = -\sqrt{-\Delta P/2l\mu_0}(2/3)r^{3/2} + \text{constant}$$

When $r = a$, $u = 0$ and constant $= \sqrt{\Delta P/2l\mu_0}(2/3)a^{3/2}$.

Hence
$$u = (2/3)\sqrt{-\Delta P/2l\mu_0}(a^{3/2} - r^{3/2})$$

$$\text{Total volumetric flow} = \int_0^a 2\pi r\,dru$$

$$= (4\pi/3)\sqrt{-\Delta P/2l\mu_0}\int_0^a (a^{3/2}r - r^{5/2})\,dr$$

$$= \underline{\underline{(2\pi/7)a^{3.5}\sqrt{-\Delta P/2l\mu_0}}}$$

Problem 3.19

Calculate the pressure drop when 3 kg/s of sulphuric acid flows through 60 m of 25 mm pipe ($\rho = 1840$ kg/m^3, $\mu = 0\cdot025$ N s/m^2).

Solution

Reynolds number $= \rho ud/\mu = 4G/\pi\mu d$
$$= 4 \times 3\cdot0/(\pi \times 0\cdot025 \times 0\cdot025)$$
$$= 6110$$

If e is taken to be $0\cdot05$ mm, $e/d = 0\cdot05/25 = 0\cdot002$.

From Fig. 3.7, $R/\rho u^2 = 0\cdot0046$.

Acid velocity in pipe $= 3\cdot0/[1840 \times (\pi/4)(0\cdot025)^2] = 3\cdot32$ m/s.

From equation 3.15, the pressure drop due to friction is given by:

$$\Delta P = 4(R/\rho u^2)(l/d)\rho u^2$$
$$= 4 \times 0.0046(60/0.025)1840 \times 3.32^2$$
$$= 89,520 \text{ N/m}^2$$
$$\simeq \underline{\underline{90 \text{ kN/m}^2}}$$

Problem 3.20

Calculate the power required to pump oil of specific gravity 0.85 and viscosity 3 mN s/m² at 4000 cm³/s through a 50 mm pipeline 100 m long, the outlet of which is 15 m higher than the inlet. The efficiency of the pump is 50%. What effect does the nature of the surface of the pipe have on the resistance?

Solution

Cross-sectional area of pipe = $(\pi/4)0.05^2 = 0.00196 \text{ m}^2$.
Velocity of oil in the pipe = $4000 \times 10^{-6}/0.00196 = 2.04$ m/s.
Reynolds number = $\rho u d/\mu = 0.85 \times 1000 \times 2.04 \times 0.05/3 \times 10^{-3}$
$$= 2.89 \times 10^4$$
If the pipe roughness e is taken to be 0.05 mm, $e/d = 0.001$, and from Fig. 3.7, $R/\rho u^2 = 0.0031$.
Equation 3.17 gives the head loss due to friction h_f as:

$$h_f = 8(R/\rho u^2)(l/d)(u^2/2g)$$
$$= 8 \times 0.0031(100/0.05)(2.04^2/2 \times 9.81)$$
$$= 10.5 \text{ m}$$

The total head = $10.5 + 15 = 25.5$ m
The mass flowrate = $4000 \times 10^{-6} \times 850 = 3.4$ kg/s
Power required = $25.5 \times 3.4 \times 9.81/0.5$
$$= 1700 \text{ W}$$
$$= \underline{\underline{1.7 \text{ kW}}}$$

The roughness of the pipe affects the ratio e/d; the rougher the pipe surface, the higher will be e/d and there will be an increase in $R/\rho u^2$. This will increase the head loss due to friction and will ultimately increase the power required.

Problem 3.21

600 cm³/s of water at 320 K is pumped in a 40 mm i.d. pipe through a length of 150 m in a horizontal direction and up through a vertical height of 10 m. In the pipe there is a control valve which may be taken as equivalent to 200 pipe diameters and other pipe fittings equivalent to 60 pipe diameters. Also in the line there is a heat

exchanger across which there is a loss in head of 1·5 m of water. If the main pipe has a roughness of 0·0002 m, what power must be delivered to the pump if the unit is 60% efficient?

Solution

Mass flowrate of water $= 600 \times 10^{-6} \times 1000 = 0.6$ kg/s.
Cross-sectional area of pipe $= (\pi/4)0.04^2 = 0.00126$ m^2.
Velocity of water in the pipe $= 600 \times 10^{-6}/0.00126 = 0.476$ m/s.
Reynolds number $= \rho u d/\mu = 1000 \times 0.476 \times 0.04/1 \times 10^{-3}$
$$= 1.9 \times 10^4$$

If $e = 0.0002$ m, $e/d = 0.005$, and from Fig. 3.7, $R/\rho u^2 = 0.0042$.
The valve and fittings are equivalent to 260 pipe diameters which is equal to $260 \times 0.04 = 10.4$ m of pipe.
The equivalent length of pipe is therefore $150 + 10.4 = 160.4$ m.
The head loss due to friction is given by equation 3.17

$$h_f = 8(R/\rho u^2)(l/d)(u^2/2g)$$
$$= 8 \times 0.0042(160.4/0.04)(0.476^2/2 \times 9.81)$$
$$= 1.56 \text{ m}$$

Hence the total head $= 1.56 + 1.5 + 10 = 13.06$ m.
The power required $= 13.06 \times 0.6 \times 9.81/0.6$
$$= \underline{\underline{128 \text{ W}}}$$

Problem 3.22

A pump developing a pressure of 800 kN/m^2 is used to pump water through a 150 mm pipe 300 m long to a reservoir 60 m higher. With the valves fully open, the flowrate obtained is 0·05 m^3/s. As a result of corrosion and scaling the effective absolute roughness of the pipe surface increases by a factor of 10. By what percentage is the flowrate reduced?

$$\text{Viscosity of water} = 1 \text{ mN s/m}^2$$

Solution

If the pump develops 800 kN/m^2, this is equivalent to a head of $(80,000/1000 \times 9.81) = 81.55$ m of water. If the pump is required to raise the water through a height of 60 m, then neglecting kinetic energy losses, the head loss due to friction in the pipe $= 81.55 - 60 = 21.55$ m.
The flowrate under these conditions is 0·05 m^3/s.
The cross-sectional area of the pipe $= (\pi/4)0.15^2 = 0.0177$ m^2.
Velocity of the water $= 0.05/0.0177 = 2.82$ m/s.

From equation 3.17, the head loss due to friction h_f is given as:

$$h_f = 8(R/\rho u^2)(l/d)(u^2/2g)$$

and

$$21 \cdot 55 = 8(R/\rho u^2)(300/0 \cdot 15)(2 \cdot 82^2/2 \times 9 \cdot 81)$$

from which

$$R/\rho u^2 = 0 \cdot 0033$$

The Reynolds number $= \rho u d/\mu = 1000 \times 2 \cdot 82 \times 0 \cdot 15/10^{-3}$
$$= 4 \cdot 23 \times 10^5$$

From Fig. 3.7, e/d is found to equal $0 \cdot 003$.

If, as a result of scaling and fouling, the roughness increases by a factor of 10, the new $e/d = 0 \cdot 03$. Fig. 3.7 can no longer be used since the new velocity, and hence the Reynolds number, is unknown. Use is made of equation 3.20 and Fig. 3.9 to find the new velocity.

The maximum head loss due to friction is still equal to $21 \cdot 55$ m as the pump head is unchanged:

$$21 \cdot 55 \text{ m} = 21 \cdot 55 \times 1000 \times 9 \cdot 81 = 211,410 \text{ N/m}^2$$

From equation 3.20,

$$(R/\rho u^2)Re^2 = \Delta P_f d^3 \rho/4l\mu^2$$
$$= 211,410 \times 0 \cdot 15^3 \times 1000/4 \times 300 \times 10^{-6}$$
$$= 6 \cdot 0 \times 10^8$$

From Fig. 3.8, $Re = 2 \cdot 95 \times 10^5$ when $e/d = 0 \cdot 03$.
Hence the new velocity $= 2 \cdot 95 \times 10^5 \times 10^{-3}/1000 \times 0 \cdot 15$
$$= 1 \cdot 97 \text{ m/s}$$
Reduction in flow $= 100(2 \cdot 82 - 1 \cdot 97)/2 \cdot 82 = \underline{\underline{30 \cdot 1\%}}$

Problem 3.23

The relation between cost per unit length C of a pipeline installation and its diameter d is given by:

$$C = a + bd$$

where a and b are independent of pipe size. Annual charges are a fraction β of the capital cost. Obtain an expression for the optimum pipe diameter on a minimum cost basis for a fluid of density ρ and viscosity μ flowing at a mass rate of G. Assume that the fluid is in turbulent flow and that the Blasius equation is applicable, i.e. the friction factor is proportional to the Reynolds number to the power of minus one quarter. Indicate clearly how the optimum diameter depends on flowrate and fluid properties.

Solution

The total annual cost of a pipeline consists of a capital charge plus the running costs. The chief element of the running cost will be the power required to overcome the head loss which is given by equation 3.17:

$$h_f = 8(R/\rho u^2)(l/d)(u^2/2g)$$

If $R/\rho u^2 = 0 \cdot 04/Re^{0 \cdot 25}$, the head loss per unit length l is given by:

$$h_f/l = 8(0 \cdot 04/Re^{0 \cdot 25})(1/d)(u^2/2g)$$
$$= 0 \cdot 016(u^2/d)(\mu/\rho ud)^{0 \cdot 25}$$
$$= 0 \cdot 016 u^{1 \cdot 75} \mu^{0 \cdot 25}/(\rho^{0 \cdot 25} d^{1 \cdot 25})$$

Now the velocity $u = G/\rho A = G/\rho(\pi/4)d^2 = 1 \cdot 27 G/\rho d^2$

$$\therefore \qquad h_f/l = 0 \cdot 016(1 \cdot 27 G/\rho d^2)^{1 \cdot 75} \mu^{0 \cdot 25}/(\rho^{0 \cdot 25} d^{1 \cdot 25})$$
$$= 0 \cdot 024 G^{1 \cdot 75} \mu^{0 \cdot 25}/(\rho^2 d^{4 \cdot 75})$$

The power required for pumping if the pump efficiency is η is:

$$P = Gg(0 \cdot 024 G^{1 \cdot 75} \mu^{0 \cdot 25}/\rho^2 d^{4 \cdot 75})/\eta$$

If $\eta = 0 \cdot 5$ (say) $P = 0 \cdot 47 G^{2 \cdot 75} \mu^{0 \cdot 25}/(\rho^2 d^{4 \cdot 75})$ (W)

If c = power cost/W, the cost of pumping is given by:

$$0 \cdot 47 c G^{2 \cdot 75} \mu^{0 \cdot 25}/\rho^2 d^{4 \cdot 75}$$

The total annual cost is then $= (\beta a + \beta bd) + (\gamma G^{2 \cdot 75} \mu^{0 \cdot 25} \mu^{0 \cdot 25}/\rho^2 d^{4 \cdot 75})$
where $\gamma = 0 \cdot 47 c$
Differentiating the total cost with respect to the diameter gives:

$$dC/dd = \beta b - 4 \cdot 75 \gamma G^{2 \cdot 75} \mu^{0 \cdot 25}/\rho^2 d^{5 \cdot 75}$$

For minimum cost, $dC/dd = 0$ and,

$$d^{5 \cdot 75} = 4 \cdot 75 \gamma G^{2 \cdot 75} \mu^{0 \cdot 25}/\rho^2 \beta b$$

and
$$d = \underline{KG^{0 \cdot 48} \mu^{0 \cdot 043}/\rho^{0 \cdot 35}}$$

where
$$K = (4 \cdot 75 \gamma/\beta b)^{0 \cdot 174}$$

Problem 3.24

A heat exchanger is to consist of a number of tubes each 25 mm diameter and 5 m long arranged in parallel. The exchanger is to be used as a cooler with a rating of 4 MW and the temperature rise in the water feed to the tubes is to be 20 K.

If the pressure drop over the tubes is not to exceed $2 \, kN/m^2$, calculate the minimum number of tubes that are required. Assume that the tube walls are smooth and that entrance and exit effects can be neglected.

Viscosity of water $= 1 \, mN \, s/m^2$.

Solution

Heat load = mass flow × specific heat × temperature rise, i.e. $4000 = m \times 4 \cdot 18 \times 20$ from which the mass flow rate $= 47 \cdot 8 \, kg/s$.

The pressure drop = $2 \text{ kN/m}^2 = 2000/1000 \times 9.81 = 0.204 \text{ m}$ of water.

Use may now be made of equation 3.20 and Fig. 3.8 to calculate the velocity in the tubes.

From equation 3.20, $(R/\rho u^2)Re^2 = \Delta P_f d^3 \rho / 4l\mu^2$

$$= 2000 \times 0.025^3 \times 1000/4 \times 5 \times 10^{-6}$$
$$= 1.56 \times 10^6$$

If the tubes are smooth, Fig. 3.8 gives a value of $Re = 2.1 \times 10^4$.

Hence the water velocity = $2.1 \times 10^4 \times 10^{-3}/1000 \times 0.025$
$$= 0.84 \text{ m/s}$$

Cross-sectional area of each tube = $(\pi/4)0.025^2 = 0.00049 \text{ m}^2$.

Mass flow rate per tube = $0.84 \times 0.00049 = 0.000412 \text{ m}^3/\text{s}$
$$= 0.412 \text{ kg/s}$$

Hence the number of tubes to meet the given specifications = $47.8/0.412$
$$= \underline{\underline{116 \text{ tubes}}}$$

Problem 3.25

Sulphuric acid is pumped at 3 kg/s through a 60 m length of smooth 25 mm pipe. Calculate the drop in pressure. If the pressure drop falls by one half, what will the new flowrate be?

Density of acid   1840 kg/m³

Viscosity of acid   25 mN s/m²

Solution

Cross-sectional area of pipe = $(\pi/4)0.025^2 = 0.00049 \text{ m}^2$.

Volumetric flowrate of acid = $3.0/1840 = 0.00163 \text{ m}^3/\text{s}$.

Velocity of acid in the pipe = $0.00163/0.00049 = 3.32 \text{ m/s}$.

Reynolds number = $\rho u d/\mu = 1840 \times 332 \times 0.025/25 \times 10^{-3}$
$$= 6120$$

From Fig. 3.7 for a smooth pipe when $Re = 6120$, $R/\rho u^2 = 0.0043$.

The pressure drop is calculated from equation 3.15:

$$\Delta P_f = 4(R/\rho u^2)(l/d)(\rho u^2)$$
$$= 4 \times 0.0043(60/0.025)(1840 \times 3.32^2)$$
$$= 837,200 \text{ N/m}^2$$
$$\approx \underline{\underline{840 \text{ kN/m}^2}}$$

If the pressure drop falls to 418,600 N/m², equation 3.20 and Fig. 3.8 may be used to calculate the new flow.

From equation 3.20, $(R/\rho u^2)Re^2 = \Delta P_f d^3 \rho / 4l\mu^2$

$$= 418,600 \times 0.025^3 \times 1840/(4 \times 60 \times 25^2 \times 10^{-6})$$
$$= 8.02 \times 10^4$$

From Fig. 3.8, $Re = 3800$ and the new velocity is given by:

$$u' = 3800 \times 25 \times 10^{-3}/1840 \times 0.025$$
$$= 2.06 \text{ m/s}$$

New mass flowrate $= 2.06 \times 0.00049 \times 1840$
$$= \underline{\underline{1.86 \text{ kg/s}}}$$

Problem 3.26

A Bingham plastic material is flowing under streamline conditions in a pipe of circular cross-section. What are the conditions for one half of the total flow to be within the central core across which the velocity profile is flat? The shear stress acting within the fluid R_y varies with velocity gradient du_x/dy according to the relation:

$$R_y - R_c = -k(du_x/dy)$$

where R_c and k are constants for the material.

Solution

The shearing characteristics of non-Newtonian fluids are shown in Fig. 3.4 where a Bingham plastic is represented by line B. This type of fluid remains rigid when the shear stress is less than the yield stress R_c and flows like a Newtonian fluid when the shear stress exceeds R_c. Examples of Bingham plastics are many fine suspensions and pastes including sewage sludge and toothpaste. The velocity profile in laminar flow is shown in Fig. 3d.

A force balance over the pipe assuming no slip at the walls gives:

$$\Delta P \pi r^2 = R_w 2\pi r L$$

and
$$\Delta P/L = 2R_w/r \qquad (1)$$

where R_w = shear stress at the wall.

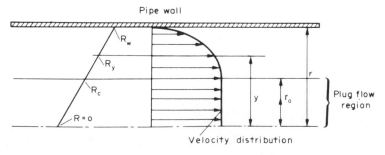

FIG. 3d

A force balance over the annular core where $y > r_0$ gives:

$$\Delta P \pi y^2 = 2\pi y L R y$$

Hence $\qquad\qquad R_y = y R_w / r \quad \text{and} \quad y = r R_y / R_w \qquad\qquad$ (2)

when $\qquad\qquad R_y = R_c \quad \text{and} \quad r_0 = r R_c / R_w \qquad\qquad$ (3)

Now $\qquad\qquad R_y - R_c = -k (du_x / dy)$

$\therefore \qquad -\dfrac{du_x}{dy} = \dfrac{R_y - R_c}{k} = \dfrac{1}{k}\left(\dfrac{y R_w}{r} - R_c\right) \quad \text{from (2)} \qquad$ (4)

Integrating, $\qquad\qquad -ku_x = (y^2 R_w / 2r) - R_c y + C$

when $y = r$, $u_x = 0$, and $C = (-r R_w / 2) + R_c r$.

$\therefore \qquad\qquad ku_x = R_w\left(\dfrac{r}{2} - \dfrac{y^2}{2r}\right) - R_c(r - y) \qquad\qquad$ (5)

Substituting for y from (3) gives:

$$ku_0 = R_w\left[\dfrac{r}{2} - \left(\dfrac{R_c}{R_w}\right)\dfrac{r}{2}\right] - R_c\left(r - \dfrac{R_c}{R_w} r\right)$$

and $\qquad\qquad ku_0 = \dfrac{r}{2R_w}(R_w - R_c)^2 \qquad\qquad$ (6)

The total volumetric flowrate Q is obtained by integrating the equation for the velocity profile as:

$$Q_{\text{total}} = \int_0^r \pi y^2(-du_x / dy)\, dy$$

$$= \frac{1}{k} \int_0^r \pi y^2\left(\frac{y R_w}{r} - R_c\right) dy \quad \text{from (4)}$$

$$= \frac{\pi r^3}{k}\left(\frac{R_w}{4} - \frac{R_c}{3}\right) \qquad (\text{m}^3/\text{s})$$

Over the central core, the volumetric flowrate Q_{core} is given by:

$$Q_{\text{core}} = \pi r_0^2 u_0 = \pi (r R_c / R_w)^2 u_0 \quad \text{from (3)}$$
$$= \pi (r R_c / R_w)^2 (r / 2k R_w)(R_w - R_c)^2 \quad \text{from (6)}$$
$$= (\pi r^3 R_c^2 / 2k R_w^3)(R_w - R_c)^2$$

If half the total flow is to be within the central core,

$$Q_{\text{core}} = Q_{\text{total}}/2$$

and $(\pi r^3 R_c^2 / 2k R_w^3)(R_w - R_c)^2 = (\pi r^3 / 2k)(R_w / 4 - R_c / 3)$

and $\qquad\qquad R_c^2(R_w - R_c)^2 = R_w^3\left(\dfrac{R_w}{4} - \dfrac{R_c}{3}\right)$

COMPRESSIBLE FLOW

Problem 4.1

Town gas, having a molecular weight of 13 kg/kmol and a kinematic viscosity of 0·25 cm²/s, is flowing through a pipe 0·25 m internal diameter and 5 km long at the rate of 0·4 m³/s and is delivered at atmospheric pressure. Calculate the pressure required to maintain this rate of flow.

The volume occupied by 1 kmol at 289 K and 101·3 kN/m² may be taken as 24·0 m³.

What effect on the pressure required would result if the gas was delivered at a height of 150 m (i) above, and (ii) below its point of entry into the pipe?

Solution

Use is made of equation 4.10 to solve this problem, and as a first approximation the kinetic energy term will be omitted:

$$(P_2 - P_1)/v_m + 4(R/\rho u^2)(l/d)(G/A)^2 = 0$$

At atmospheric pressure and 289 K, the density = 13/24·0 = 0·542 kg/m³.
Mass flowrate of gas. $G = 0\cdot4 \times 0\cdot542 = 0\cdot217$ kg/s.
Cross-sectional area $A = (\pi/4)(0\cdot25)^2 = 0\cdot0491$ m².
Gas velocity $u = 0\cdot4/0\cdot0491 = 8\cdot146$ m/s

$$\therefore \qquad G/A = 0\cdot217/0\cdot0491 = 4\cdot413 \text{ kg/m}^2\text{s}$$

Reynolds number, $Re = \rho du/\mu$

$$= 0\cdot25 \times 8\cdot146/0\cdot25 \times 10^{-4}$$
$$= 8\cdot146 \times 10^4$$

For $e/d = 0\cdot002$, $R/\rho u^2 = 0\cdot0031$ from Fig. 3.7,

$$v_2 = 1/0\cdot542 = 1\cdot845 \text{ m}^3/\text{kg}$$
$$v_1 = (22\cdot4/13)(298/273)(101\cdot3/P_1)$$
$$= 190\cdot5/P_1$$
$$v_m = (0\cdot923P_1 + 95\cdot25)/P_1 \quad \text{m}^3/\text{kg}$$

Substituting in equation 4.10 gives

$$P_1(P_1 - 101\cdot3)10^3/(0\cdot923P_1 + 95\cdot25) = 4(0\cdot0031)(5000/0\cdot25)(4\cdot726)^2$$

from which $\qquad P_1 = \underline{111 \cdot 1 \text{ kN/m}^2}$

The kinetic energy term $= (G/A)^2 \ln (P_1/P_2) = (4 \cdot 413)^2 \ln (111 \cdot 1/101 \cdot 3)$
$$= 1 \cdot 81 \text{ kg}^2/\text{m}^4\text{s}^2$$

This is negligible in comparison with the other terms which equal 5539 kg^2/m^4 s^2 so that the initial approximation is justified. If the pipe is not horizontal, the term $g\,dz$ in equation 4.4 must be included in the calculation. If equation 4.4 is divided throughout by v^2, on integration this term becomes:

$$g\Delta z/v_m{}^2$$

$\therefore \qquad\qquad v_m = (0 \cdot 923 \times 111 \cdot 1 + 95 \cdot 25)/111 \cdot 1 = 1 \cdot 781 \text{ m}^3/\text{kg}$

$\qquad\qquad\quad v_{\text{air}} = (24 \cdot 0/29) = 0 \cdot 827 \text{ m}^3/\text{kg}$

As gas is less dense than air, v_m is replaced by $v_{\text{air}} - v_m = -0 \cdot 954 \text{ m}^3/\text{kg}$.

Hence $\qquad\qquad g\Delta z/v_m{}^2 = 9 \cdot 81 \times 150/0 \cdot 954^2$
$$= 1616 \text{ N/m}^2$$
$$\equiv 0 \cdot 16 \text{ kN/m}^2$$

(i) If the delivery point is 150 m above the entry level, then since gas is less dense,
$$P_1 = 111 \cdot 1 - 0 \cdot 16 = \underline{110 \cdot 94 \text{ kN/m}^2}$$

(ii) If the delivery point is 150 m below the entry level then,
$$P_1 = 111 \cdot 1 + 0 \cdot 16 = \underline{111 \cdot 26 \text{ kN/m}^2}$$

Problem 4.2

Nitrogen at 12 MN/m^2 pressure is fed through a 25 mm diameter mild steel pipe to a synthetic ammonia plant at the rate of 1·25 kg/s. What will be the drop in pressure over a 30 m length of pipe for isothermal flow of the gas at 298 K?

Absolute roughness of the pipe surface	= 0·005 mm
Kilogram molecular volume	= 22·4 m^3
Viscosity of nitrogen	= 0·02 mN s/m^2

Solution

Molecular weight of nitrogen = 28 kg/kmol.
Assume a mean pressure in the pipe = 10 MN/m^2.
Specific volume, v_m at 10 MN/m^2 and 298 K is given by:

$$v_m = (22 \cdot 4/28)(101 \cdot 3/10 \times 10^3)(298/273)$$
$$= 0 \cdot 00885 \text{ m}^3/\text{kmol}.$$

Reynolds number $= \rho u d/\mu = d(G/A)\mu$.
$A = (\pi/4)(0 \cdot 025)^2 = 4 \cdot 91 \times 10^{-3} \text{ m}^2$.

Hence $G/A = 1.25/4.91 \times 10^{-3} = 2540 \text{ kg/m}^2 \text{ s}$

and
$$Re = 0.025 \times 2540/0.02 \times 10^{-3}$$
$$= 3.18 \times 10^6$$

From Fig. 3.7, when $Re = 3.18 \times 10^6$ and $e/d = 0.005/25 = 0.0002$,

$$R/\rho u^2 = 0.0017$$

Use is now made of equation 4.10, neglecting the first term,

i.e.
$$(P_2 - P_1)/v_m + 4(R/\rho u^2)(l/d)(G/A)^2 = 0$$

or
$$P_1 - P_2 = 4v_m(R/\rho u^2)(l/d)(G/A)^2$$
$$= 4 \times 0.00885(0.0017)(30/0.025)(2540)^2$$
$$= 466,000 \text{ N/m}^2$$
$$\equiv 0.466 \text{ MN/m}^2$$

This is small in comparison with $P_1 = 12 \text{ MN/m}^2$, and the average pressure of 10 MN/m^2 is seen to be too low. Therefore a new mean pressure of 11.75 kN/m^2 is selected and the above calculation repeated to give a new pressure drop of 0.39 MN/m^2. The mean pressure is therefore $(12 + 11.61)/2 = 11.8 \text{ MN/m}^2$ which is near enough to the assumed value.

It remains to check whether the assumption that the kinetic energy term was negligible was justified:

$$KE \text{ term} = (G/A)^2 \ln (P_1/P_2)$$
$$= (2540)^2 \ln (12/11.61)$$
$$= 2.13 \times 10^5 \text{ kg}^2/\text{m}^4\text{s}^2$$

The term $(P_1 - P_2)/v_m$ where v_m is the specific volume at the mean pressure of $11.75 \text{ MN/m}^2 = 0.39 \times 10^6/0.00753$.

$$= 5.18 \times 10^7 \text{ kg}^2/\text{m}^4\text{s}$$

Hence it is justifiable to omit the kinetic energy term and

the pressure drop $= \underline{\underline{0.39 \text{ MN/m}^2}}$

Problem 4.3

Hydrogen is pumped from a reservoir at 2 MN/m^2 pressure through a clean horizontal mild steel pipe 50 mm diameter and 500 m long. The downstream pressure is also 2 MN/m^2 and the pressure of this gas is raised to 2.6 MN/m^2 by a pump at the upstream end of the pipe. The conditions of flow are isothermal and the temperature of the gas is 293 K. What is the flowrate and what is the effective rate of working of the pump?

Viscosity of hydrogen $= 0.009 \text{ mN s/m}^2$ at 293 K

Solution

Use is made of equation 4.8 to solve this problem and in the first instance the kinetic energy term will be neglected:

$$(P_2^2 - P_1^2)/2P_1v_1 + 4(R/\rho u^2)(l/d)(G/A)^2 = 0$$

where $P_1 = 2 \cdot 6$ MN/m², $P_2 = 2 \cdot 0$ MN/m²

$$v_1 = (22 \cdot 4/2)(293/273)(0 \cdot 1013/2 \cdot 6) = 0 \cdot 468 \text{ m}^3/\text{kg}$$

A Reynolds number must be assumed to obtain a value of $R/\rho u^2$.
If $Re = 10^7$ and $e/d = 0 \cdot 001$, $R/\rho u^2 = 0 \cdot 0023$
Substituting in the above equation gives:

$$(2 \cdot 0^2 - 2 \cdot 6^2)10^{12}/(2 \times 2 \cdot 6 \times 10^6 \times 0 \cdot 468) + 4(0 \cdot 0023)(500/0 \cdot 05)(G/A)^2 = 0$$

from which $G/A = 111$ kg/m²s.

$$Re = d(G/A)/\mu = 0 \cdot 05 \times 111/(0 \cdot 009 \times 10^{-3}) = 6 \cdot 2 \times 10^5$$

Thus Re was chosen too high. If Re is taken as $6 \cdot 0 \times 10^5$ and the problem reworked, $G/A = 108$ kg/m²s and $Re = 6 \cdot 03 \times 10^5$ which gives good agreement.

$$A = (\pi/4)(0 \cdot 05)^2 = 0 \cdot 00197 \text{ m}^2$$

Hence
$$G = 108 \times 0 \cdot 00197 = \underline{0 \cdot 213 \text{ kg/s}}$$

The power requirement is given by equation 6.57, i.e.

$$\text{power} = (1/\eta)GP_1v_1 \ln (P_1/P_2)$$

If a 60% efficiency is assumed, then,

$$\text{power} = (1/0 \cdot 6) \times 0 \cdot 213 \times 2 \cdot 6 \times 10^6 \times 0 \cdot 468 \ln (2 \cdot 6/2)$$
$$= 1 \cdot 13 \times 10^5 \text{ W} \equiv \underline{\underline{113 \text{ kW}}}$$

Problem 4.4

In a synthetic ammonia plant the hydrogen is fed through a 50 mm steel pipe to the converters. The pressure drop over the 30 m length of pipe is 500 kN/m², the pressure at the downstream end being $7 \cdot 5$ MN/m². What power is required in order to overcome friction losses in the pipe? Assume isothermal expansion of the gas at 298 K. What error is introduced by assuming the gas to be an incompressible fluid of density equal to that at the mean pressure in the pipe? The gas viscosity = $0 \cdot 02$ mN s/m².

Solution

If the downstream pressure = $7 \cdot 5$ MN/m² and the pressure drop due to friction =

500 kN/m^2, the upstream pressure $= 8 \cdot 0 \text{ MN/m}^2$ and the mean pressure $= 7 \cdot 75 \text{ MN/m}^2$.

The mean specific volume is given by:

$$v_m = (22 \cdot 4/2)(298/273)(0 \cdot 1013/7 \cdot 75)$$
$$= 0 \cdot 16 \text{ m}^3/\text{kg}$$

and
$$v_1 = (22 \cdot 4/2)(298/273)(0 \cdot 1013/8 \cdot 0) = 0 \cdot 15 \text{ m}^3/\text{kg}$$

It is necessary to assume a value of $R/\rho u^2$, calculate G/A and the Reynolds number and check that the value of e/d is reasonable. If the gas is assumed to be an incompressible fluid of density equal to the mean pressure in the pipe, and taking a value of $R/\rho u^2 = 0 \cdot 003$, the pressure drop due to friction $= 500 \text{ kN/m}^2$:

$$500 \times 10^3/0 \cdot 16 = 4(0 \cdot 003)(30/0 \cdot 05)(G/A)^2$$

and
$$G/A = 658 \text{ kg/m}^2\text{s}.$$

$$Re = d(G/A)/\mu = 0 \cdot 05 \times 658/0 \cdot 02 \times 10^{-3} = 1 \cdot 65 \times 10^6$$

From Fig. 3.7 this corresponds to a value of e/d of approximately $0 \cdot 002$, which is reasonable for a steel pipe.

For compressible flow use is made of equation 4.8:

$$(G/A)^2 \ln (P_1/P_2) + (P_2^2 - P_1^2)/2P_1 v_1 + 4(R/\rho u^2)(l/d)(G/A)^2 = 0$$

Substituting gives:

$$(G/A)^2 \ln (8 \cdot 0/7 \cdot 5) + (7 \cdot 5^2 - 8 \cdot 0^2)10^{12}/(2 \times 8 \cdot 0 \times 10^6 \times 0 \cdot 15)$$
$$+ 4(0 \cdot 003)(30/0 \cdot 05)(G/A)^2 = 0$$

from which
$$G/A = 667 \text{ kg/m}^2\text{s}$$

and
$$G = 667 \times (\pi/4)(0 \cdot 05)^2 = 1 \cdot 31 \text{ kg/s}$$

Very little error is made by the simplifying assumption in this particular case. The power requirement is given by equation 6.57:

$$\text{power} = (1/\eta)GP_1 v_1 \ln (P_1/P_2)$$

If the compressor efficiency $= 60\%$,

$$\text{power} = (1/0 \cdot 6) \times 1 \cdot 31 \times 8 \cdot 0 \times 10^6 \times 0 \cdot 15 \ln (8/7 \cdot 5)$$
$$= 1 \cdot 69 \times 10^5 \text{ W} \equiv \underline{\underline{169 \text{ kW}}}$$

Problem 4.5

A vacuum distillation plant operating at 7 kN/m^2 pressure at the top has a boil-up rate of $0 \cdot 125 \text{ kg/s}$ of xylene. Calculate the pressure drop along a 150 mm bore vapour pipe used to connect the column to the condenser.

Solution

From vapour pressure date, the vapour temperature $= 338$ K and the molecular weight of xylene $= 106$ kg/kmol.

Use is made of equation 4.8 to solve this problem, i.e.

$$(G/A)^2 \ln (P_1/P_2) + (P_2^2 - P_1^2)/2P_1v_1 + 4(R/\rho u^2)(l/d)(G/A)^2 = 0$$

Cross-sectional area of pipe, $A = (\pi/4)(0.15)^2 = 1.76 \times 10^{-2}$ m^2
$$G/A = 0.125/1.76 \times 10^{-2} = 7.07 \text{ kg/m}^2\text{s}$$

The Reynolds number is given by $\rho u d/\mu = d(G/A)/\mu$
$$= 0.15 \times 7.07/(0.01 \times 10^{-3})$$
$$= 1.06 \times 10^5$$

From Fig. 3.7 for $e/d = 0.002$ and $Re = 1.06 \times 10^5$, $(R/\rho u^2) = 0.003$.

Specific volume $v_1 = (22.4/106)(338/273)(101.3/7.0) = 3.79$ m^3/kg.

Substituting in equation 4.8 gives:

$$(7.07)^2 \ln (7/P_2) + (P_2^2 - 7^2) \times 10^6/2 \times 7 \times 10^3 \times 3.79 + 4 \times 0.003(6/0.15)(7.07)^2 = 0$$

where $P_2 = $ pressure at the condenser (kN/m^2).

This equation is solved by trial and error to give:

$$P_2 = 6.91 \text{ kN/m}^2$$

Hence
$$P_1 - P_2 = 7.0 - 6.91 = 0.09 \text{ kN/m}^2$$
$$\equiv 90 \text{ N/m}^2$$

Problem 4.6

Nitrogen at 12 MN/m^2 pressure is fed through a 25 mm diameter mild steel pipe to a synthetic ammonia plant at the rate of 0.4 kg/s. What will be the drop in pressure over a 30 m length of pipe assuming isothermal expansion of the gas at 300 K? What is the average quantity of heat per unit area of pipe surface that must pass through the walls in order to maintain isothermal conditions? What would be the pressure drop in the pipe if it were perfectly lagged?

$$\mu = 0.02 \text{ mN s/m}^2$$

Solution

At the high pressure quoted in this problem the kinetic energy term in equation 4.8 may be neglected to give:

$$(P_2^2 - P_1^2)/2P_1v_1 + 4(R/\rho u^2)(l/d)(G/A)^2 = 0$$

Specific volume at entry of pipe $v_1 = (22.4/28)(300/273)(0.1013/12)$
$$= 0.00742 \text{ m}^3/\text{kg}$$

Cross-sectional area of pipe $A = (\pi/4)(0.025)^2$
$$= 0.00049 \text{ m}^2$$

Hence $G/A = 0.4/0.00049 = 816 \text{ kg/m}^2\text{s}$.

Reynolds number, $d(G/A)/\mu = 0.025 \times 816/(0.02 \times 10^{-3})$
$$= 1.02 \times 10^6$$

If $e/d = 0.002$ and $Re = 1.02 \times 10^6$, $R/\rho u^2 = 0.0028$ from Fig. 3.7.

Substituting gives:

$$(12^2 - P_2{}^2)10^{12}/(2 \times 12 \times 10^6 \times 0.00742) = 4(0.0028)(30/0.025)(816)^2$$

from which $P_2 = 11.93 \text{ MN/m}^2$

and the pressure drop $= (12.0 - 11.93) = 0.07 \text{ MN/m}^2 \equiv \underline{\underline{70 \text{ kN/m}^2}}$

The heat required to maintain isothermal flow is shown in Section 4.2.1 to be $G\Delta u^2/2$.

The velocity at the high pressure end of the pipe

$$= \text{volumetric flow/area}$$

$$= (G/A)v_1 = (816 \times 0.0072) = 6.06 \text{ m/s}$$

and the velocity in the plant is taken as zero:

$$\therefore \qquad G\,\Delta u^2/2 = 0.4 \times (6.06)^2/2 = 7.34 \text{ W}$$

Outside area of pipe $= 30 \times \pi \times 0.025 = 2.36 \text{ m}^2$.

Heat required $= 7.34/2.36 = \underline{\underline{3.12 \text{ W/m}^2}}$

This low value of the heat required is because the change in kinetic energy is small and conditions are almost adiabatic. If the pipe were perfectly lagged, the flow would be adiabatic and the pressure drop would then be calculated from equations 4.29 and 4.24. Equation 4.29 enables the specific volume at the low pressure end v_2 to be calculated from:

$$8(R/\rho u^2)(l/d) = \left[\frac{\gamma-1}{2\gamma} + \frac{P_1}{v_1}\left(\frac{A}{G}\right)^2\right]\left[1 - \left(\frac{v_1}{v_2}\right)^2\right] - \frac{\gamma+1}{\gamma}\ln\left(\frac{v_2}{v_1}\right)$$

For nitrogen, $\gamma = 1.4$ and substitution gives:

$$8(0.0028)(30/0.025) = \left[\frac{1.4-1}{2\times1.4} + \frac{12\times10^6}{0.00742}\left(\frac{1}{816}\right)^2\right]\left[1 - \left(\frac{0.00742}{v_2}\right)^2\right]$$
$$- \frac{1.4+1}{1.4}\ln\left(\frac{v_2}{0.00742}\right)$$

from which by trial and error, $v_2 = 0.00746 \text{ m}^3\text{/kg}$.

This value of v_2 is now substituted into equation 4.24 to find P_2:

$$\frac{1}{2}\left(\frac{G}{A}\right)^2 v_1{}^2 + \frac{\gamma}{\gamma-1}P_1 v_1 = \frac{1}{2}\left(\frac{G}{A}\right)^2 v^2 + \frac{\gamma}{\gamma-1}P_2 v_2$$

Substituting $\qquad (816)^2(0.00742)^2/2 + [1.4/(1.4-1)]12 \times 10^6 \times 0.00742$

$$= (816)^2(0.00746)^2/2 + [1.4/(1.4-1)]P_2 \times 10^6 \times 0.00746$$

from which $P_2 = 11.94 \text{ MN/m}^2$

and the pressure drop for adiabatic flow $= 12.0 - 11.94$
$$= 0.06 \text{ MN/m}^2$$
$$\equiv \underline{\underline{60 \text{ kN/m}^2}}$$

Problem 4.7

Air, at a pressure of 10 MN/m² and a temperature of 290 K, flows from a reservoir through a mild steel pipe of 10 mm diameter and 30 m long into a second reservoir at a pressure P_2. Plot the mass rate of flow of the air as a function of the pressure P_2. Neglect any effects attributable to differences in level and assume an adiabatic expansion of the air.

$$\mu = 0\cdot018 \text{ mN s/m}^2, \ \gamma = 1\cdot36$$

Solution

G/A is required as a function of P_2. v_2 cannot be found directly since the downstream temperature T_2 is unknown and varies as a function of the flowrate. For adiabatic flow, use may be made of equation 4.29 to calculate v_2 using specified values of G/A. This value of v_2 may then be substituted in equation 4.24 to obtain the value of P_2. In this way the required data may be calculated.

Equation 4.29 states:

$$8(R/\rho u^2)(l/d) = \left[\frac{\gamma-1}{2\gamma} + \frac{P_1}{v_1}\left(\frac{A}{G}\right)^2\right]\left[1 - \left(\frac{v_1}{v_2}\right)^2\right] - \frac{\gamma+1}{\gamma}\ln\left(\frac{v_2}{v_1}\right)$$

and equation 4.24 is:

$$0\cdot5(G/A)^2v_1^2 + [\gamma/(\gamma-1)]P_1v_1 = 0\cdot5(G/A)^2v_2^2 + [\gamma/(\gamma-1)]P_2v_2$$

or

$$\frac{0\cdot5(G/A)^2(v_1^2 - v_2^2) + [\gamma/(\gamma-1)]P_1v_1}{[\gamma/(\gamma-1)]v_2} = P_2$$

When $P_2 = P_1 = 10$ MN/m², $G/A = 0$.
If G/A is specified as 2000 kg/m²s,

$$Re = 0\cdot01 \times 2000/0\cdot018 \times 10^{-3}$$
$$= 1\cdot11 \times 10^6$$

From Fig. 3.7 when $e/d = 0\cdot002$, $R/\rho u^2 = 0\cdot0028$.

$$v_1 = (22\cdot4/29)(290/273)(0\cdot1013/10)$$
$$= 0\cdot0083 \text{ m}^3 \text{ kg}$$

Substituting into equation 4.29 gives:

$$8(0\cdot0028)(30/0\cdot01) = \left[\frac{0\cdot36}{2 \times 1\cdot36} + \frac{10 \times 10^6}{0\cdot0083}\left(\frac{1}{2000}\right)^2\right]\left[1 - \left(\frac{0\cdot0083}{v_2}\right)^2\right] - \frac{2\cdot36}{1\cdot36}\ln\left(\frac{v_2}{0\cdot0083}\right)$$

from which $v_2 = 0\cdot00942$ m³/kg.
Substituting for v_2 in equation 4.24 gives:

$$P_2 = [0\cdot5(2000)^2(0\cdot0083^2 - 0\cdot00942^2) + (1\cdot36/0\cdot36)10$$
$$\times 10^6 \times 0\cdot0083]/(1\cdot36/0\cdot36) \times 0\cdot00942$$

and $P_2 = 8\cdot75$ MN/m².

In a similar way the following table may be produced.

G/A kg/m²s	v_2 m³/kg	P_2 MN/m²
0	0·0083	10·0
2000	0·00942	8·75
3000	0·012	6·76
3500	0·0165	5·01
4000	0·025	3·37
4238	0·039	2·04

These data are plotted in Fig. 4a.

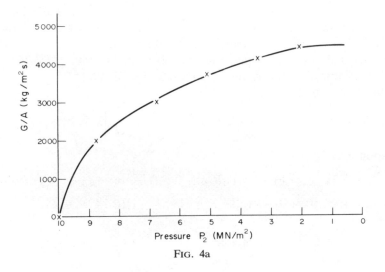

FIG. 4a

It is shown in Section 4.2.3 that the maximum velocity which can occur in a pipe under adiabatic flow conditions is sonic velocity which is equal to $\sqrt{\gamma P_2 v_2}$.

From the above table $\sqrt{\gamma P_2 v_2}$ at maximum flow is given by

$$\sqrt{1\cdot36 \times 2\cdot04 \times 10^6 \times 0\cdot039} = 329 \text{ m/s}$$

The temperature at this condition is given by $P_2 v_2 = \mathbf{R}T/M$, so that

$$T_2 = 29 \times 0\cdot039 \times 2\cdot04 \times 10^6/8314$$
$$= 277 \text{ K}$$

The velocity of sound at 277 K = 334 m/s which serves as a check on the calculated data.

Problem 4.8

In order to reduce transport costs it has been suggested that gasworks should be located in the coal-mining areas and the gas be distributed to the centres of population through high-pressure mains. Discuss the engineering and economic problems which are likely to be involved in such a project. If a trunk pipeline is 200 mm diameter and 80 km long, derive an expression for the rate of flow of coal gas in terms of the upstream and downstream pressures P_1 and P_2 respectively, on the assumption that the gas temperature remains at 290 K and that the ideal gas law is followed. Calculate the maximum rate of flow when the upstream pressure is 10 MN/m². What is the pressure at the downstream end of the pipe under these conditions?

Viscosity of coal gas	$= 0{\cdot}02$ mN s/m²
Mean molecular weight	$= 13$ kg/kmol
Molecular volume	$= 22{\cdot}4$ m³/kmol
Pipe roughness	$= 0{\cdot}05$ mm

Solution

It is beyond the scope of a problems book of this type to provide a full answer to the first part of this question. Points which should be considered, however, would include, under the heading of engineering problems, aspects of high-pressure design, valves, controls, seals, etc. Economic considerations would include the problem of scale of manufacture on a smaller number of larger plants and increased pipe and pumping costs. Additional points might include the social problems caused by such a change, new distribution areas, and the security such a network might involve.

Equation 4.8 gives the relationship between P_1, P_2, and G/A, i.e.

$$(G/A)^2 \ln(P_1/P_2) + (P_2^2 - P_1^2)/2P_1 v_1 + 4(R/\rho u^2)(l/d)(G/A)^2 = 0$$

The derivation of this equation is discussed in Section 4.2.

The maximum rate of flow and the downstream pressure under these conditions, P_w, may be calculated from equations 4.15 and 4.16.

Equations 4.16 expresses the relationship between P_w and P_1 by

$$\ln(P_1/P_w)^2 + 1 - (P_1/P_w)^2 + 8(R/\rho u^2)(l/d) = 0$$
$$e/d = 0{\cdot}05/200 = 0{\cdot}00025$$

As the Reynolds number is unknown, assume $R/\rho u^2 = 0{\cdot}00175$. Substituting gives:

$$\ln(10/P_w)^2 + 1 - (10/P_w)^2 + 8(0{\cdot}00175)(80 \times 10^3/0{\cdot}2) = 0$$

from which $P_w = 0{\cdot}134$ MN/m².

Specific volume $v_1 = (22{\cdot}4/13)(290/273)(0{\cdot}1013/10) = 0{\cdot}0185$ m³/kg.

Equation 4.15 gives the maximum flow rate as

$$G_w = A P_w \sqrt{1/(P_1 v_1)}$$

$$= (\pi/4)(0{\cdot}2)^2 \times 0{\cdot}134 \times 10^6 \times \sqrt{1/(10 \times 10^6 \times 0{\cdot}0185)}$$

$$= 9{\cdot}78 \text{ kg/s}$$

It remains to check that the assumed value of $R/\rho u^2$ was satisfactory:

$$A = (\pi/4)(0 \cdot 2)^2 = 0 \cdot 0314 \text{ m}^2$$
$$G/A = 9 \cdot 78/0 \cdot 0314 = 311 \text{ kg/s m}^2$$
$$Re = 0 \cdot 2 \times 311/(0 \cdot 02 \times 10^{-3}) = 3 \cdot 11 \times 10^6$$

From Fig. 3.7, $R/\rho u^2 = 0 \cdot 00175$, which agrees with the assumed value. Therefore, the maximum flow $= \underline{9 \cdot 78 \text{ kg/s}}$

and
$$P_w = \underline{\underline{0 \cdot 134 \text{ MN/m}^2}}$$

Problem 4.9

Over a 30 m length of 150 mm vacuum line carrying air at 293 K the pressure falls from 1 kN/m² to 0·1 kN/m². If the relative roughness e/d is 0·002, what is the approximate flowrate?

Solution

Specific volume of air at 293 K and 1 kN/m² is given by:

$$v_1 = (22 \cdot 4/29)(293/273)(101 \cdot 3/1 \cdot 0)$$
$$= 83 \cdot 98 \text{ m}^3/\text{kg}$$

It is necessary to assume a Reynolds number to determine $R/\rho u^2$ and then calculate G/A, the value of which should correspond to the original assumed value. Assume the Reynolds number $= 1 \times 10^5$.
From Fig. 3.7 for $e/d = 0 \cdot 002$ and $Re = 10^5$, $R/\rho u^2 = 0 \cdot 003$.
Use is now made of equation 4.8, i.e.

$$(G/A)^2 \ln(P_1/P_2) + (P_2^2 - P_1^2)/2P_1 v_1 + 4(R/\rho u^2)(l/d)(G/A)^2 = 0$$

Substituting gives:

$$(G/A)^2 \ln(1 \cdot 0/0 \cdot 1) + (0 \cdot 1^2 - 1^2) \times 10^6/(2 \times 1 \times 10^3 \times 83 \cdot 98)$$
$$+ 4(0 \cdot 003)(30/0 \cdot 15)(G/A)^2 = 0$$

from which $(G/A) = 1 \cdot 37 \text{ kg/m}^2\text{s}$.
The viscosity of air is 0·018 mN s/m² so that,

$$Re = 0 \cdot 15 \times 1 \cdot 37/(0 \cdot 018 \times 10^{-3}) = 1 \cdot 14 \times 10^4$$

Thus the chosen value of Re was too high. Choosing $Re = 1 \times 10^4$ gives $R/\rho u^2 = 0 \cdot 0041$ and $G/A = 1 \cdot 26 \text{ kg/m}^2\text{s}$.
If $G/A = 1 \cdot 26 \text{ kg/m}^2\text{s}$, Re now equals $1 \cdot 04 \times 10^4$ which is good agreement:

$$G/A = 1 \cdot 26 \text{ kg/m}^2\text{s}$$
$$\therefore \qquad G = 1 \cdot 26 \times (\pi/4) \times (0 \cdot 15)^2 = \underline{\underline{0 \cdot 022 \text{ kg/s}}}$$

Problem 4.10

A vacuum system is required to handle 10 g/s of vapour (molecular weight 56 kg/kmol) so as to maintain a pressure of 1·5 kN/m² in a vessel situated 30 m from the vacuum pump. If the pump is able to maintain a pressure of 0·15 kN/m² at its suction point, what diameter pipe is required? The temperature is 290 K, and isothermal conditions may be assumed in the pipe, whose surface can be taken as smooth. The ideal gas law is followed.

$$\text{Gas viscosity} = 0\cdot01 \text{ mN s/m}^2$$

Solution

Use is made of equation 4.8 to solve this problem. It is necessary to assume a value of the pipe diameter d to calculate values of G/A, Reynolds number and $R/\rho u^2$.
Choose $d = 0\cdot10$ m giving $A = (\pi/4)(0\cdot10)^2 = 0\cdot00785$ m²

$\therefore$
$$G/A = 10 \times 10^{-3}/0\cdot00785 = 1\cdot274 \text{ kg/m}^2\text{s}$$

and
$$Re = d(G/A)/\mu = 0\cdot10 \times 1\cdot274/(0\cdot01 \times 10^{-3})$$
$$= 1\cdot274 \times 10^4$$

For a smooth pipe, $R/\rho u^2 = 0\cdot0035$ from Fig. 3.7.
Specific volume at inlet $v_1 = (22\cdot4/56)(290/273)(101\cdot3/1\cdot5)$
$$= 28.7 \text{ m}^3/\text{kg}$$

Equation 4.8 gives:

$$(G/A)^2 \ln(P_1/P_2) + (P_2^2 - P_1^2)/2P_1v_1 + 4(R/\rho u^2)(l/d)(G/A)^2 = 0$$

Substituting gives:

$$(1\cdot274)^2 \ln(1\cdot5/0\cdot15) + (0\cdot15^2 - 1\cdot5^2) \times 10^6/(2 \times 1\cdot5 \times 10^3 \times 28.7)$$
$$+ (0\cdot0035)(30/0\cdot10)(1\cdot274)^2 = -16\cdot3$$

Therefore the chosen value of d was too large.
A second assumed of $d = 0\cdot05$ m gives a value of equation 4.8 = 25·9 and the procedure is repeated until the right-hand side of equation 4·8 = 0.
This occurs when $d = 0\cdot08$ m $\equiv$ 80 mm.

Problem 4.11

In a vacuum system, air is flowing isothermally at 290 K through a 150 mm diameter pipeline 30 m long. If the relative roughness of the pipewall e/d is 0·002 and the downstream pressure is 130 N/m², what will the upstream pressure be if the flowrate of air is 0·025 kg/s?
Assume that the ideal gas law applies and that the viscosity of air is constant at 0·018 mN s/m².
What error would be introduced if the change in kinetic energy of the gas as a result of expansion were neglected?

Solution

As the upstream and mean specific volumes v_1 and v_m are required in equations 4.8 and 4.10 respectively, use is made of equation 4.11 to solve this problem:

$$(G/A)^2 \ln(P_1/P_2) + (P_2^2 - P_1^2)/(2RT/M) + 4(R/\rho u^2)(l/d)(G/A)^2 = 0$$

$$R = 8 \cdot 314 \, \text{kJ/kmol K}$$

$$\therefore \qquad 2RT/M = 2 \times 8 \cdot 314 \times 10^3 \times 290/29 = 1 \cdot 66 \times 10^5 \, \text{J/kg}$$

NB—The middle term has units of $(N/m^2)^2/(J/kg) = kg^2/s^2 \, m^4$ which are consistent with those of the other terms.

$$A = (\pi/4)(0 \cdot 15)^2 = 0 \cdot 0176 \, \text{m}^2$$

$$\therefore \qquad G/A = 0 \cdot 025/0 \cdot 0176 = 1 \cdot 414$$

and $\qquad Re = d(G/A)/\mu = 0 \cdot 15 \times 1 \cdot 414/(0 \cdot 018 \times 10^{-3}) = 1 \cdot 18 \times 10^4$

For Fig. 3.7 for smooth pipes and $Re = 1 \cdot 18 \times 10^4$, $R/\rho u^2 = 0 \cdot 0040$. Substituting in equation 4.11 gives:

$$(1 \cdot 414)^2 \ln(P_1/130) + (130^2 - P_1^2)/1 \cdot 66 \times 10^5 + 4 \times 0 \cdot 0040(30/0 \cdot 15)(1 \cdot 414)^2 = 0$$

A trial and error solution gives the upstream pressure $P_1 = \underline{\underline{1 \cdot 36 \, \text{kN/m}^2}}$.

If the kinetic energy term were neglected, equation 4.11 becomes:

$$(P_2^2 - P_1^2)/(2RT/M) + 4(R/\rho u^2)(l/d)(G/A)^2 = 0$$

and the solution to this equation is $P_1 = \underline{\underline{1 \cdot 04 \, \text{kN/m}^2}}$.

Thus considerable error would be introduced by this simplifying assumption.

Problem 4.12

Air is flowing at a rate of 30 kg/m²s through a smooth pipe of 50 mm diameter and 300 m long. If the upstream pressure is 800 kN/m², what will the downstream pressure be if the flow is isothermal at 273 K? Take the viscosity of air as 0·015 mN s/m² and the kg molecular volume as 22·4 m³. What is the significance of the change in kinetic energy of the fluid?

Solution

Use is made of equation 4.8 to solve this problem:

$$(G/A)^2 \ln(P_1/P_2) + (P_2^2 - P_1^2)/2P_1 v_1 + 4(R/\rho u^2)(l/d)(G/A)^2 = 0$$

The specific volume at the upstream condition v_1 is given by:

$$v_1 = (22 \cdot 4/29)(273/273)(101 \cdot 3/800)$$

$$= 0 \cdot 098 \, \text{m}^3/\text{kg}$$

$$G/A = 30 \, \text{kg/m}^2 \, \text{s}$$

$$Re = 0.05 \times 30/(0.015 \times 10^{-3})$$
$$= 1.0 \times 10^5$$

For a smooth pipe, $R/\rho u^2 = 0.0032$ from Fig. 3.7.
Hence substituting in equation 4.8 gives:

$$(30)^2 \ln(800/P_2) + (P_2^2 - 800^2) \times 10^6/(2 \times 800 \times 10^3 \times 0.098)$$
$$+ 4(0.0032)(300/0.05)(30)^2 = 0$$

from which the downstream pressure $P_2 = \underline{793 \text{ kN/m}^2}$.

The kinetic energy term $= (G/A)^2 \ln(800/793)$
$$= 7.91 \text{ kg}^2/\text{m}^4 \text{ s}$$

This is insignificant in comparison with 69,120 kg²/m⁴ s which is the value of the other terms in equation 4.8.

Problem 4.13

If temperature does not change with height, estimate the boiling point of water at a height of 3000 m above sea level. The barometer reading at sea level is 98·4 kN/m² and the temperature is 288·7 K. The vapour pressure of water at 288·7 K is 1·77 kN/m². The molecular weight of air is 29 kg/kmol.

Solution

If the air pressure at 3000 m $= P_2$ kN/m², the pressure at sea level $P_1 = 98.4$ kN/m² then:

$$\int v \, dP + \int g \, dz = 0$$

Now
$$v = v_1(P/P_1)$$

so that
$$P_1 v_1 \int \frac{dP}{P} + \int g \, dz = 0$$

and
$$P_1 v_1 \ln(P_2/P_1) + g(z_2 - z_1) = 0$$

In this problem
$$v_1 = (22.4/29)(288.7/273)(101.3/98.4)$$
$$= 0.841 \text{ m}^3/\text{kg}.$$

Hence
$$98400 \times 0.841 \ln(P_2/98.4) + 9.81(3000 - 0) = 0$$

and
$$P_2 = 68.95 \text{ kN/m}^2$$

The relationship between vapour pressure and temperature may be expressed as:

$$\log P = a + bT$$

When
$$T = 288.7, \quad P = 1.77 \text{ kN/m}^2$$

and when
$$T = 373, \quad P = 101.3 \text{ kN/m}^2$$

from which
$$\log P = -5.773 + 0.0209T$$

Now $P_2 = 68.95$ so that T is given by the above equation as $\underline{364 \text{ K}}$.

Problem 4.14

A 150 mm gas main is used for transferring coal gas (molecular weight 13 kg/kmol and kinematic viscosity $0 \cdot 25$ cm²/s) at 295 K from a plant to a storage station 100 m away, at a rate of 1 m³/s. Calculate the pressure drop if the pipe can be considered to be smooth.

If the maximum permissible pressure drop is 10 kN/m², is it possible to increase the flowrate by 25%?

Solution

If the flow of 1 m³/s is at STP, the specific volume of the gas at STP is given by $(22 \cdot 4/13) = 1 \cdot 723$ m³/kg.

The mass flowrate, $G = 1 \cdot 0/1 \cdot 723 = 0 \cdot 58$ kg/s.

Cross-sectional area, $A = (\pi/4)(0 \cdot 15)^2 = 0 \cdot 0176$ m²

$\therefore \qquad\qquad\qquad G/A = 32 \cdot 82$ kg/m²s

$$\mu/\rho = 0 \cdot 25 \text{ cm}^2/\text{s} = 0 \cdot 25 \times 10^{-4} \text{ m}^2/\text{s}$$

and $\qquad\qquad \mu = 0 \cdot 25 \times 10^{-4} \times (1/1 \cdot 723)$

$$= 1 \cdot 45 \times 10^{-5} \text{ N s/m}^2$$

$\therefore \qquad\qquad\qquad Re = 0 \cdot 15 \times 32 \cdot 82/1 \cdot 45 \times 10^{-5}$

$$= 3 \cdot 4 \times 10^5$$

From Fig. 3.7, for smooth pipes, $R/\rho u^2 = 0 \cdot 0017$.

The pressure drop due to friction is calculated as follows:

$$4(R/\rho u^2)(l/d)(G/A)^2 = 4(0 \cdot 0017)(100/0 \cdot 15)(32 \cdot 82)^2$$

$$= 4883 \text{ kg}^2/\text{m}^4 \text{ s}^2$$

and $\Delta P = (4883/1 \cdot 723) = 2834$ N/m² $\equiv \underline{\underline{2 \cdot 83 \text{ kN/m}^2}}$.

If the flow is increased by 25%,

$$G = 1 \cdot 25 \times 0 \cdot 58 = 0 \cdot 725 \text{ kg/s}$$

$$G/A = 41 \cdot 19$$

$$Re = 0 \cdot 15 \times 41 \cdot 9/1 \cdot 45 \times 10^5$$

$$= 4 \cdot 3 \times 10^5$$

and $\qquad\qquad R/\rho u^2 = 0 \cdot 00165$ from Fig. 3.17.

The new pressure drop $= 4(0 \cdot 00165)(100/0 \cdot 15)(41 \cdot 19)^2 1 \cdot 723$

$$= \underline{\underline{4 \cdot 33 \text{ kN/m}^2}} \text{ (which is less than 10 kN/M}^2\text{)}$$

It is therefore possible to measure the flowrate by 25%.

SECTION 5

FLOW MEASUREMENT

Problem 5.1

Sulphuric acid of specific gravity 1·3 is flowing through a pipe of 50 mm internal diameter. A thin-lipped orifice, 10 mm diameter, is fitted in the pipe and the differential pressure shown by a mercury manometer is 10 cm. Assuming that the leads to the manometer are filled with the acid, calculate (a) the weight of acid flowing per second, and (b) the approximate loss of pressure (in kN/m^2) caused by the orifice.

The coefficient of discharge of the orifice may be taken as 0·61, the specific gravity of mercury as 13·55, and the density of water as 1000 kg/m^3.

Solution

(a) The mass flowrate G is given by equations 5.17 and 5.19,

i.e.

$$G = \frac{C_D A_0}{v} \sqrt{\frac{2v(P_1 - P_2)}{1 - (A_0/A_1)^2}}$$

or, where $[1 - (A_0/A_1)^2]$ is approximately unity,

$$G = C_D A_0 \rho \sqrt{(2gh)}$$

where h is the difference in head across the orifice expressed in terms of the fluid in question.

In this problem, $A_0 = (\pi/4)(0 \cdot 01)^2 = 7 \cdot 85 \times 10^{-5} \, m^2$

$$A_1 = (\pi/4)(0 \cdot 05)^2 = 196 \cdot 3 \times 10^{-5} \, m^2$$

and

$$[1 - (A_0/A_1)^2] = [1 - (7 \cdot 85/196 \cdot 3)^2] = 0 \cdot 999$$

$$h = 10 \text{ cm mercury}$$

$$= (10/1000)(13 \cdot 55 - 1 \cdot 0)/1 \cdot 0 \quad \text{(from equation 5.3)}$$

$$= 0 \cdot 094 \text{ m of sulphuric acid}$$

Substituting gives:

$$G = 0 \cdot 61 \times 7 \cdot 85 \times 10^{-5} \times (1 \cdot 3 \times 1000) \sqrt{(2 \times 9 \cdot 81 \times 0 \cdot 094)}$$

$$= 0 \cdot 085 \text{ kg/s}$$

(b) The drop in pressure $= \rho gh$

$$= 1300 \times 9\cdot81 \times 0\cdot094$$
$$= 1198 \text{ N/m}^2$$
$$\equiv 1\cdot2 \text{ kN/m}^2$$

Problem 5.2

The rate of discharge of water from a tank is measured by means of a notch, for which the flowrate is directly proportional to the height of liquid above the bottom of the notch. Calculate and plot the profile of the notch if the flowrate is $0\cdot01$ m³/s when the liquid level is 150 mm above the bottom of the notch.

Solution

The velocity of a fluid particle discharged at a height h above any level $= \sqrt{2gh}$ and the velocity will therefore normally vary from the top to the bottom of a notch. Consider now an horizontal element of fluid of length $2l$ and thickness δh flowing through a notch at a height h from the bottom of the notch.

The velocity of the fluid through the strip $= \sqrt{2gh}$.

The area of the fluid element $= 2l\delta h$.

Volumetric flow through the strip $=$ velocity $\times$ area

$$= \sqrt{2gh}\, 2l\delta h$$

Integrating from the bottom to the top of the notch will give the total discharge Q,

i.e.
$$Q = 2l\sqrt{2g} \int_0^H \sqrt{h}\; dh$$

$$= 2l\sqrt{2g}\, H^{3/2}$$

The problem states that the flow Q is directly proportional to H, so that

$$Q = 2l\sqrt{2g}\, H^{3/2} = KH$$

where K is a proportionality constant.

When $Q = 0\cdot01$ m³/s, $H = 0\cdot15$ m, so that:

$$0\cdot01 = K \times 0\cdot15 \quad \text{and} \quad K = 0\cdot0667 \text{ m}^2/\text{s}$$

$$\therefore \qquad 2l\sqrt{2g}\sqrt{H} = 0\cdot0667$$

or
$$H = (0\cdot0667)^2/(4l^2 \times 2g)$$
$$= 0\cdot0000566/l^2$$

If l is expressed in mm and H in m, then:

$$H = 56\cdot6/l^2$$

Thus a table can be produced giving the required profile of the notch.

l (mm)	H (m)	l (mm)	H (m)
10	0·566	50	0·023
15	0·251	75	0·010
20	0·142	100	0·006
25	0·091		

The required profile is plotted in Fig. 5a .

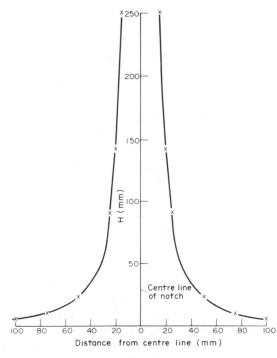

FIG. 5a

Problem 5.3

Water flows at between 3000 and 4000 cm³/s through a 50 mm pipe and is metered by means of an orifice. Suggest a suitable size of orifice if the pressure difference is to be measured with a simple water manometer. What approximately is the pressure difference recorded at the maximum flow rate?

Solution

Equations 5.17 and 5.19 relate the pressure drop to the mass flowrate. If equation

5.19 is first used as an approximation,

$$G = C_D A_0 \rho \sqrt{(2gh)}$$

Considering the maximum flow of 4000 cm³/s, $G = 4$ kg/s. The largest practicable height of a water manometer will be taken as 1 m and equation 5.19 used to calculate the orifice area A_0. The coefficient of discharge C_D will be taken as 0·6

Then $4·0 = 0·6 A_0 \times 1000 \sqrt{(2 \times 9·81 \times 1·0)}$

                            $A_0 = 0·0015$ m²

and $d_0 = 0·0438$ m

This diameter d_0 is comparable with the pipe diameter and $[1 - (A_0/A_1)^2] = [1 - (43·8^2/50^2)^2] = 0·641$. Therefore the value of A_0 must be recalculated as:

$$4·0 = 0·6 A_0 \times 1000 \sqrt{(2 \times 9·8 \times 1·0)/[1 - (A_0/A_1)^2]}$$

from which $A_0 = 0·00195$ m²

and $d = 0·039$ m $= \underline{39 \text{ mm}}$

Now $\sqrt{[1 - (A_0/A_1)^2]} = \sqrt{[1 - (39^2/50^2)^2]} = 0·793$

Substituting in equation 5.17 gives:

$$4·0 = 0·6 \times 0·00195 \times 1000 \sqrt{(2 \times 0·001 \Delta P / 0·793)}$$

from which $\Delta P = 12320$ N/m²

                            $\equiv \underline{\underline{12·3 \text{ kN/m}^2}}$

Problem 5.4

The rate of flow of water in a 150 mm diameter pipe is measured with a venturi meter with a 50 mm diameter throat. When the pressure drop over the converging section is 100 mm of water, the flowrate is 2·7 kg/s. What is the coefficient for the converging cone of the meter at that flowrate and what is the head lost due to friction? If the total loss of head over the meter is 15 mm water, what is the coefficient for the diverging cone?

Solution

The equation relating the mass flowrate G and the head loss across a venturi meter is given by equations 5.17, 5.32 and 5.33:

$$G = \frac{C_D A_2}{v} \sqrt{\frac{2v(P_1 - P_2)}{1 - (A_2/A_1)^2}} \qquad \text{(equation 5.17)}$$

$$= C_D \rho \frac{A_1 A_2}{\sqrt{(A_1^2 - A_1^2)}} \sqrt{(2v(P_1 - P_2))} \qquad \text{(equation 5.32)}$$

$$= C_D \rho C' \sqrt{(2gh_v)} \qquad \text{(equation 5.33)}$$

where C' is a constant for the meter and h_v is the loss in head over the converging cone expressed as height of fluid.

$$A_1 = (\pi/4)(0\cdot15)^2 = 0\cdot0176\ \text{m}^2$$
$$A_2 = (\pi/4)(0\cdot05)^2 = 0\cdot00196\ \text{m}^2$$
$$C' = 0\cdot0176 \times 0\cdot00196/\sqrt{(0\cdot0176^2 - 0\cdot00196^2)}$$
$$= 0\cdot00197\ \text{m}^2$$
$$h_v = 0\cdot1\ \text{m}$$
$$\therefore \qquad 2\cdot7 = C_D \times 1000 \times 0\cdot00197\sqrt{(2 \times 9\cdot81 \times 0\cdot10)}$$
$$\text{and} \qquad \underline{\underline{C_D = 0\cdot978}}$$

Considering equation 5.33, if there were no losses the coefficient of discharge of the meter would be unity, and for a flowrate G the loss in head would be $h_v - h_f$ where h_f is the head loss due to friction.

Then
$$G = \rho C'\sqrt{[2g(h_v - h_f)]}$$

Dividing this equation by equation 5.33 and squaring gives:

$$1 - (h_f/h_v) = C_D^2$$
$$\text{and} \qquad h_f = h_v(1 - C_D^2)$$

Hence
$$h_f = 100(1 - 0\cdot978^2)$$
$$= \underline{\underline{4\cdot35\ \text{mm}}}$$

If the head recovered over the diverging cone is h_v' and the coefficient of discharge for the converging cone is C_D', then:

$$G = C_D'\rho C'\sqrt{(2gh_v')}$$

If the whole of the excess kinetic energy is recovered as pressure energy, the coefficient C_D' will equal unity and G will be obtained with a recovery of head equal to h_v' plus some quantity h_f',

$$G = \rho C'\sqrt{[2g(h_v' + h_f')]}$$

Equating these last two equations and squaring gives:

$$C_D'^2 = 1 + (h_f'/h_v')$$
$$\text{and} \qquad h_f' = h_v'(C_D'^2 - 1)$$

Thus the coefficient of the diverging cone is greater than unity and the total loss of head $= h_f + h_f'$.

Head loss over diverging cone $= 15\cdot0 - 4\cdot35$
$$= 10\cdot65\ \text{mm}$$

The coefficient of the diverging cone C_D' is given by:

$$G = C_D'\rho C'\sqrt{(2gh_v')}$$
$$h_v' = 100 - 15 = 85\ \text{mm}$$

and $2 \cdot 7 = C_D' \times 1000 \times 0 \cdot 00197 \sqrt{(2 \times 9 \cdot 81 \times 0 \cdot 085)}$

from which $$\underline{\underline{C_D' = 1 \cdot 06}}$$

Problem 5.5

A venturi meter with a 50 mm throat is used to measure a flow of slightly salt water in a pipe of inside diameter 100 mm. The meter is checked by adding 20 cm^3/s of normal sodium chloride solution above the meter and analysing a sample of water downstream from the meter. Before addition of the salt, 1000 cm^3 of water requires 10 cm^3 of 0·1 M silver nitrate solution in a titration. 1000 cm^3 of the downstream sample required 23·5 cm^3 0·1 M silver nitrate. If a mercury-under-water manometer connected to the meter gives a reading of 205 mm, what is the discharge coefficient of the meter? Assume that the density of the liquid is not appreciably affected by the salt.

Solution

It is necessary to determine the flowrate of the slightly salt water from a mass balance on the salt. The salt concentrations may be obtained from the titration readings, noting that normality N is equal to molarity in this case.

For the upstream solution: $N_u \times 1000 = 0 \cdot 1 \times 10$

$$N_u = 0 \cdot 001 = 0 \cdot 001 \times 58 \cdot 5 \text{ g/l}$$

$$\equiv 58 \cdot 5 \times 10^{-6} \text{ g/cm}^3 \text{ NaCl}$$

For the downstream solution: $N_d \times 1000 = 0 \cdot 1 \times 23 \cdot 5$

$$N_d = 0 \cdot 00235 = 0 \cdot 00235 \times 58 \cdot 5 \text{ g/l}$$

$$\equiv 137 \cdot 5 \times 10^{-6} \text{ g/cm}^3 \text{ NaCl}$$

A salt balance gives for a flowrate of x cm^3/s,

$$x \times 58 \cdot 5 \times 10^{-6} + 20 \times 58 \cdot 5/1000 = 137 \cdot 5 \times 10^{-6}(x + 20)$$

from which $x = 14780 \text{ cm}^3/\text{s}$

$$\equiv 0 \cdot 0148 \text{ m}^3/\text{s or } 14 \cdot 8 \text{ kg/s}$$

$$= 14 \cdot 8 \text{ kg/s}$$

For the venturi meter: throat area $A_1 = (\pi/4)(0 \cdot 05)^2 = 0 \cdot 00196 \text{ m}^2$

pipe area $A_2 = (\pi/4)(0 \cdot 10)^2 = 0 \cdot 00785 \text{ m}^2$

Using equations 5.32 and 5.33, C_D may be determined:

$$C' = A_1 A_2 / \sqrt{(A_1^2 - A_2^2)}$$

$$= 0 \cdot 00204 \text{ m}^2$$

$$h = 205 \text{ mmHg} = (0 \cdot 205 \times 13 \cdot 55/1)$$

$$= 2 \cdot 78 \text{ m of water}$$

∴ $14 \cdot 8 = C_D \times 1000 \times 0 \cdot 00204 \sqrt{(2 \times 9 \cdot 81 \times 2 \cdot 78)}$

and $$\underline{\underline{C_D = 0 \cdot 982}}$$

Problem 5.6

A gas cylinder containing $30 \, \text{m}^3$ of air at $6 \, \text{MN/m}^2$ pressure discharges to the atmosphere through a valve which may be taken as equivalent to a sharp edged orifice of 6 mm diameter (coefficient of discharge = 0·6). Plot the rate of discharge against the pressure in the cylinder. How long will it take for the pressure in the cylinder to fall to (a) $1 \, \text{MN/m}^2$, and (b) $150 \, \text{kN/m}^2$?

Assume an adiabatic expansion of the gas through the valve and that the contents of the cylinder remain at 273 K.

Solution

Area of orifice $= (\pi/4)(0 \cdot 006)^2 = 2 \cdot 828 \times 10^{-5} \, \text{m}^2$.
The critical pressure ratio w_c is given by equation 5.28:

$$w_c = [2/(k + 1)]^{k/(k-1)}$$

Taking $k = \gamma = 1 \cdot 4$ for air, $w_c = 0 \cdot 527$.
Thus sonic velocity will occur until the cylinder pressure falls to a pressure $P_2 = 101 \cdot 3/0 \cdot 527 = 192 \cdot 2 \, \text{kN/m}^2$.
For pressures in excess of $192 \cdot 2 \, \text{kN/m}^2$, the rate of discharge is given by equation 5.31:

$$G = C_D A_0 \sqrt{(kP_1/v_1)(2/(k + 1))^{(k+1)/(k-1)}}$$

If $k = 1 \cdot 4$ and the values of C_D and A_0 are substituted:

$$G = 1 \cdot 162 \times 10^{-5} \sqrt{(P_1/v_1)}$$

If P_a and v_a are atmospheric pressure and the specific volume at atmospheric pressure respectively,

$$P_a v_a = P_1 v_1 \quad \text{and} \quad v_1 = P_a v_a/P_1$$

Now $P_a = 101,300 \, \text{N/m}^2$ and $v_a = (22 \cdot 4/29) = 0 \cdot 773 \, \text{m}^3/\text{kg}$

$$v_1 = 101,300 \times 0 \cdot 773/P_1 = 78,246/P_1$$

and
$$G = 1 \cdot 162 \times 10^{-5} \sqrt{(P_1^2/78,246)}$$

$$= 4 \cdot 15 \times 10^{-8} P_1 \quad \text{kg/s}$$

If P_1 is expressed as MN/m^2, $\underline{G = 0 \cdot 0415 P_1 \quad \text{kg/s}}$.

For pressures lower than $192 \cdot 2 \, \text{kN/m}^2$, use is made of equation 5.24, which, on squaring gives:

$$G^2 = (A_0 C_D/v_2)^2 2P_1 v_1 (k/k - 1)[1 - (P_2/P_1)^{(k-1)/k}]$$

Now
$$v_2 = v_a = 0 \cdot 773 \, \text{m}^3/\text{kg}$$
$$P_2 = P_a = 101,300 \, \text{N/m}^2$$
$$v_1 = P_a v_a/P_1$$

Substituting these values gives:

$$\underline{\underline{G^2 = 2 \cdot 64 \times 10^{-4}[1 - (P_a/P_1)^{0 \cdot 286}]}}$$

Thus a table of G as a function of pressure can be produced.

$P < 192 \cdot 2 \, \text{kN/m}^2$		$P > 192 \cdot 2 \, \text{kN/m}^2$	
P (MN/m²)	G (kg/s)	P (MN/m²)	G (kg/s)
0·1013	0	0·2	0·0083
0·110	0·0024	0·5	0·0208
0·125	0·0039	1·0	0·0416
0·150	0·0053	2·0	0·0830
0·175	0·0062	6·0	0·249

This data is plotted in Fig. 5b to provide the answer to the first part of the problem. The discharge rate is seen to be linear until the cylinder pressure falls to $0 \cdot 125 \, \text{MN/m}^2$.

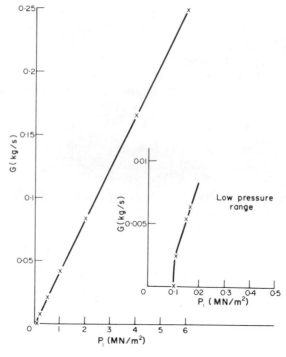

FIG. 5b

If m is the mass of air in the cylinder at any pressure P_1 over the linear part of the curve, $G = \mathrm{d}m/\mathrm{d}t = 0 \cdot 0415 P_1$.

$$\therefore \qquad \mathrm{d}t = \mathrm{d}m/0 \cdot 0415 P_1$$

Now
$$m = (29/22 \cdot 4)(P_1/0 \cdot 1013) \times 30 = 383 \cdot 4 P_1 \quad \text{kg}$$

Hence
$$\mathrm{d}t = 383 \cdot 4 \, \mathrm{d}m/0 \cdot 0415 m = 9240 \, \mathrm{d}m/m$$

$$\therefore \qquad t = 9240 \ln(m_1/m_2)$$

At 6 MN/m² and 1 MN/m² respectively, the mass of air in the cylinder is 2308 and 383·4 kg respectively.

∴ Time for pressure to fall to 1 MN/m² = 9240 ln(2308/383·4)

$$= \underline{\underline{16600 \text{ s}}}$$

As 0·15 MN/m² is still within the linear region, the time for the pressure to fall to this value is found to be 34100 s.

Problem 5.7

Air at a pressure of 1500 kN/m² and a temperature of 370 K flows through an orifice of 30 mm² to atmospheric pressure.

If the coefficient of discharge is 0·65, the critical pressure ratio 0·527, and the ratio of the specific heats is 1·4, calculate the weight flowing per second.

Solution

If the critical pressure ratio w_c from equation 5.28 is 0·527, sonic velocity will occur until the pressure falls to 101·3/0·527 = 192·2 kN/m². For pressures above this value, use is made of equation 5.31 to determine the mass flowrate,

i.e. $$G = C_D A_0 \sqrt{(kP_1/v_1)[2/(k+1)]^{(k+1)/(k-1)}}$$

If $k = 1·4$,

$$G = C_D A_0 \sqrt{(1·4P_1/v_1)(2/2·4)^{2·4/0·4}}$$
$$= C_D A_0 \sqrt{(0·468P_1/v_1)}$$

$$P_1 = 1,500,000 \text{ N/m}^2$$

and $$v_1 = (22·4/29)(370/273)(101·3/1500)$$
$$= 0·0707 \text{ m}^3/\text{kg}$$

Substituting gives $G = 0·65 \times 30 \times 10^{-6} \sqrt{(0·486 \times 1,500,000/0·0707)}$
$$= \underline{\underline{0·061 \text{ kg/s}}}$$

Problem 5.8

Water flows through an orifice of 25 mm diameter situated in a 75 mm pipe at the rate of 300 cm³/s. What will be the difference in level on a water manometer connected across the meter? Take the viscosity of water as 1 mN s/m².

Solution

Area of orifice $A_0 = (\pi/4)(0·025)^2 = 0·00049 \text{ m}^2$.
Area of pipe $A_1 = (\pi/4)(0·075)^2 = 0·0044 \text{ m}^2$.

$$[1 - (A_0/A_1)^2] = [1 - (0·00049/0·0044)^2]$$
$$= 0·994$$

Use may be made of equation 5.19 to solve this problem.
Mass flowrate $G = 300 \times 10^{-6} \times 1000$
$$= 0.30 \text{ kg/s}$$
Reynolds number in orifice $= d_0(G/A)/\mu$
$$= 0.025 \times (0.30/0.00049)/1 \times 10^{-3}$$
$$= 1.53 \times 10^{-4}$$
The ratio, (orifice diameter/pipe diameter) $= 25/75 = 0.33$.
From Fig. 5.12, the coefficient of discharge $C_D = 0.61$.
Equation 5.19 states that $G = C_D A_0 \rho \sqrt{(2gh)}$.
Substituting gives $0.30 = 0.61 \times 0.00049 \times 1000\sqrt{(2 \times 9.81h)}$
from which $\qquad h = 0.051 \text{ m}$
$$\equiv \underline{\underline{51 \text{ mm}}}$$

Problem 5.9

Water flowing at 1500 cm³/s in a 50 mm diameter pipe is metered by means of a simple orifice of diameter 25 mm. If the coefficient of discharge of the meter is 0.62, what will be the reading on a mercury-under-water manometer connected to the meter?
What is the Reynolds number for the flow in the pipe?
(Density of water $= 1000 \text{ kg/m}^3$; viscosity of water $= 1 \text{ mN s/m}^2$).

Solution

The mass flowrate, $G = 1500 \times 10^{-6} \times 1000$
$$= 1.5 \text{ kg/s}$$
Area of orifice, $A_0 = (\pi/4)(0.025)^2 = 0.00049 \text{ m}^2$.
Area of pipe, $A_1 = (\pi/4)(0.050)^2 = 0.00196 \text{ m}^2$.
Reynolds number $= \rho u d/\mu = d(G/A_1)/\mu$
$$= 0.05(1.5/0.00196)/(1 \times 10^{-3})$$
$$= \underline{\underline{3.83 \times 10^4}}$$

The orifice meter equations are 5.17 and 5.19 with the latter being used when $\sqrt{[1 - (A_0/A_1)^2]}$ approaches unity.

In this example $\qquad \sqrt{[1 - (A_0/A_1)^2]} = \sqrt{[1 - (25^2/50^2)^2]}$
$$= 0.968$$

Using equation 5.19, $\qquad G = C_D A_0 \rho \sqrt{(2gh)}$
$$1.5 = 0.62 \times 0.00049 \times 1000\sqrt{(2 \times 9.81h)}$$
and $\qquad h = 1.24 \text{ m of water}$

Using equation 5.17 in terms of h gives:

$$1.5 = (0.62 \times 0.00049 \times 1000/0.968)\sqrt{(2gh)}$$
and $\qquad h = 1.16 \text{ m of water}$

This latter value of h should be used. The height of a mercury-under-water manometer would then be $1·16/(13·55 - 1·0) = 0·092$ m $= \underline{\underline{92 \text{ mmHg}}}$.

Problem 5.10

What size of orifice would give a pressure difference of 0·3 m water gauge for the flow of a petroleum product of specific gravity 0·9 at 0·05 m^3/s in a 150 mm diameter pipe?

Solution

As in the earlier problems, equations 5.17 and 5.19 may be used to calculate the flow through an orifice. In this problem the size of the orifice is to be found so that the simpler equation 5.19 will be used in the first instance.

$$G = C_D A_0 \rho \sqrt{(2gh)}$$
$$G = 0·05 \times 0·9 \times 1000$$
$$= 45·0 \text{ kg/s}$$
$$\rho = 0·9 \times 1000 = 900 \text{ kg/m}^3$$
$$h = 0·3 \text{ m of water}$$

or
$$(0·3/0·9) = 0·333 \text{ m of petroleum product}$$
$$C_D = 0·62 \text{ (assumed)}$$

Hence
$$45·0 = 0·62 \times A_0 \times 900 \sqrt{(2 \times 9·81 \times 0·333)}$$

and
$$A_0 = 0·03155 \text{ m}^2$$

∴
$$d_0 = 0·2 \text{ m}$$

This orifice diameter is larger than the pipe size so that it was clearly wrong to use the simpler equation. Using equation 5.17 with the pressure difference expressed as head of liquid gives:

$$G = C_D A_0 \rho \sqrt{[2gh/(1 - (A_0/A_1)^2)]}$$
$$A_1 = (\pi/4)(0·15)^2 = 0·0177 \text{ m}^2$$

Hence
$$45·0 = 0·62 \times A_0 \times 900 \sqrt{[2 \times 9·81 \times 0·33/(1 - (A_0/0·0177)^2)]}$$

from which
$$A_0 = 0·0154 \text{ m}^2$$

and
$$d_0 = \underline{\underline{0·14 \text{ m}}}$$

Problem 5.11

The flow of water through a 50 mm pipe is measured by means of an orifice meter with a 40 mm aperture. The pressure drop recorded is 150 mm on a mercury-under-water manometer and the coefficient of discharge of the meter is 0·6. What is the

Reynolds number in the pipe and what would you expect the pressure drop over a 30 m length of the pipe to be?

$$\text{Friction factor, } \phi = R/\rho u^2 = 0{\cdot}0025$$
$$\text{Specific gravity of mercury } = 13{\cdot}6$$
$$\text{Viscosity of water } \qquad = 1 \text{ mN s/m}^2$$

What type of pump would you use, how would you drive it, and what material of construction would be suitable?

Solution

Area of pipe, $A_1 = (\pi/4)(0{\cdot}05)^2 = 0{\cdot}00197 \text{ m}^2$.
Area of orifice, $A_0 = (\pi/4)(0{\cdot}04)^2 = 0{\cdot}00126 \text{ m}^2$.
$h = 150$ mmHg under water $= 0{\cdot}15 \times (13{\cdot}55 - 1{\cdot}0)/1{\cdot}0 \equiv 1{\cdot}88$ m of water.
$1 - (A_0/A)^2 = 0{\cdot}591$, so that equation 5.17 must be used:

$$G = C_D A_0 \rho \sqrt{[2gh/(1 - (A_0/A)^2)]}$$
$$= 0{\cdot}6 \times 0{\cdot}00126 \times 1000\sqrt{(2 \times 9{\cdot}81 \times 1{\cdot}88/0{\cdot}591)}$$
$$= 5{\cdot}97 \text{ kg/s}$$

Reynolds number $= \rho u d/\mu = d(G/A_1)/\mu$
$$= 0{\cdot}05(6{\cdot}22/0{\cdot}00197)/(1 \times 10^{-3})$$
$$= \underline{\underline{1{\cdot}52 \times 10^5}}$$

The pressure drop is found from:

$$\Delta P/v = 4(R/\rho u^2)(l/d)(G/A)^2$$
$$= 4(0{\cdot}0025)(30/0{\cdot}05)(5{\cdot}97/0{\cdot}00197)^2$$
$$= 5{\cdot}74 \times 10^7 \text{ kg}^2/\text{m}^4 \text{ s}^2$$
$$\Delta P = 5{\cdot}74 \times 10^7 \times (1/1000)$$
$$= 5{\cdot}74 \times 10^4 \text{ N/m}^2$$
$$\equiv \underline{\underline{57{\cdot}4 \text{ kN/m}^2}}$$

Power required $=$ head loss (m) $\times G \times g$
$$= (5{\cdot}74 \times 10^4/1000 \times 9{\cdot}81) \times 5{\cdot}97 \times 9{\cdot}81$$
$$= 343 \text{ W}$$

If the pump efficiency $= 60\%$, actual power requirement $= 343/0{\cdot}6 = 571$ W.
The water velocity $= 5{\cdot}97/0{\cdot}00197 \times 1000$
$$= 3{\cdot}03 \text{ m/s}$$

For this low-power requirement at a low head and comparatively low flowrate, a centrifugal pump, electrically driven and made of stainless steel, would be employed.

Problem 5.12

A rotameter has a tube 0·3 m long which has an internal diameter of 25 mm at the top and 20 mm at the bottom. The diameter of the float is 20 mm, its effective specific gravity is 4·80, and its volume 6·6 cm³. If the coefficient of discharge is 0·72, at what height will the float be when metering water at 100 cm³/s?

Solution

Area at the top of the tube $= (\pi/4)(25)^2 = 491$ mm².
Area at the bottom of the tube $= (\pi/4)(20)^2 = 314$ mm².
Area of the float $= 314$ mm².
Volume of float $= 6\cdot6$ cm³ $\equiv 6600$ mm³.
Equation 5.36 relates the geometry of the rotameter and the fluid density to the mass flowrate G:

$$G = C_D A_2 \sqrt{\frac{2gV_f(\rho_f - \rho)\rho}{A_f(1 - (A_2/A_1)^2)}}$$

where A_1 and A_2 are the areas of the tube and the annulus respectively.

$$G = 100 \text{ cm}^3/\text{s} = 0\cdot1 \text{ kg/s}$$
$$\rho_f = 4800 \text{ kg/m}^3$$
$$\rho = 1000 \text{ kg/m}^3$$

Initially it will be assumed that $[1 - (A_2/A_1)^2]^{0\cdot5}$ is approximately equal to unity. Then,

$$0\cdot1 = 0\cdot72A_2\sqrt{[2 \times 9\cdot81 \times 6600 \times 10^{-9}(4800 - 1000)1000/3\cdot14 \times 10^{-6}]}$$

and
$$A_2 = 0\cdot000111 \text{ m}^2$$
$$\equiv 111 \text{ mm}^2$$

The outside diameter d of the annulus between the tube wall and the float is then given by:

$$111 = (\pi/4)(d^2 - 20^2)$$
and
$$d = 23\cdot27 \text{ mm}$$

The height of the top of the float from the bottom of the tube is given by h:

$$h = 0\cdot3(23\cdot27 - 20)/(25 - 20)$$
$$= 0\cdot196 \text{ m}$$

The area can now be corrected to allow for the initial assumption that $[1 - (A_2/A_1)^2]^{0\cdot5}$ was equal to unity,

$$\sqrt{[1 - (A_2/A_1)^2]} = \sqrt{[1 - (111)/(\pi/4)(23\cdot27)^2]^2}$$
$$= 0\cdot965$$
$$\therefore \qquad \text{true } A_2 = 111 \times 0\cdot965$$
$$= 107\cdot15 \text{ mm}^2$$

and $d = 23 \cdot 16 \, \text{mm}$

Float height $= 0 \cdot 3(23 \cdot 16 - 20)/(25 - 20)$
$= \underline{\underline{0 \cdot 19 \, \text{m}}}$

Problem 5.13

Explain why there is a critical pressure ratio across a nozzle at which, for a given upstream pressure, the flowrate is a maximum. Obtain an expression for the maximum flow for a given upstream pressure for isentropic flow through a horizontal nozzle. Show that for air (ratio of specific heats, $\gamma = 1 \cdot 4$) the critical pressure ratio is $0 \cdot 53$ and calculate the maximum flow through an orifice of area $30 \, \text{mm}^2$ and coefficient of discharge $0 \cdot 65$ when the upstream pressure is $1 \cdot 5 \, \text{MN/m}^2$ and the upstream temperature $293 \, \text{K}$.

(Kilogram molecular volume $= 22 \cdot 4 \, \text{m}^3$)

Solution

The first part of this problem is fully discussed in Section 5.3.3. The maximum rate of discharge is derived in that section and the equation is presented as 5.31 as:

$$G = C_D A_0 \sqrt{(kP_1/v_1)(2/k + 1)^{(k+1)/(k-1)}}$$

For an isentropic process, $k = \gamma = 1 \cdot 4$ for air.
The critical pressure ratio w_c is shown in equation 5.28 to be given by:

$$w_c = (2/k + 1)^{k/(k-1)}$$

Substituting for $k = \gamma = 1 \cdot 4$

$$w_c = (2/2 \cdot 4)^{1 \cdot 4/0 \cdot 4}$$
$$= \underline{\underline{0 \cdot 523}}$$

The maximum rate of discharge is given by equation 5.31 above.

In this example: $P_1 = 1 \cdot 5 \times 10^6 \, \text{N/m}^2$

$A_0 = 30 \times 10^{-6} \, \text{m}^2$

$k = 1 \cdot 4$

and $C_D = 0 \cdot 65$

At $P_1 = 1 \cdot 5 \, \text{MN/m}^2$ and $T_1 = 293 \, \text{K}$, the specific volume v_1 is given by:

$$v_1 = (22 \cdot 4/29)(293/273)(0 \cdot 1013/1 \cdot 5)$$
$$= 0 \cdot 056 \, \text{m}^3/\text{kg}$$

Substituting gives $G = 0 \cdot 65 \times 30 \times 10^{-6} \sqrt{(1 \cdot 4 \times 1 \cdot 5 \times 10^6/0 \cdot 056)(2/2 \cdot 4)^{2 \cdot 4/0 \cdot 4}}$
$= \underline{\underline{0 \cdot 069 \, \text{kg/s}}}$

Problem 5.14

A gas cylinder containing air discharges to atmosphere through a valve whose characteristics may be considered similar to those of a sharp-edged orifice. If the pressure in the cylinder is initially $350\ kN/m^2$, by how much will the pressure have fallen when the flowrate has decreased to one quarter of its initial value?

The flow through the valve may be taken as isentropic and the expansion in the cylinder as isothermal. The ratio of the specific heats at constant pressure and constant volume is 1·4.

Solution

From equation 5.28 the critical pressure ratio w_c is given by:

$$w_c = [2/(k+1)]^{k/(k-1)}$$
$$= (2/2 \cdot 4)^{1 \cdot 4/0 \cdot 4}$$
$$= 0 \cdot 528$$

If the cylinder is discharging to atmospheric pressure, sonic velocity will occur until the cylinder pressure has fallen to:

$$101 \cdot 3/0 \cdot 528 = 192\ kN/m^2$$

Equation 5.31 relates the maximum discharge when the cylinder pressure exceeds $192\ kN/m^2$:

$$G = C_D A_0 \sqrt{\frac{kP_1}{v_1}\left(\frac{2}{k+1}\right)^{(k+1)(k-1)}}$$

If P_a and v_a are the pressure and specific volume at atmospheric pressure, then:

$$1/v_1 = P_1/P_a v_a$$

and

$$G = C_D A_0 \sqrt{\frac{kP_1^2}{P_a v_a}\left(\frac{2}{k+1}\right)^{(k+1)/(k-1)}}$$
$$= C_D A_0 P_1 \sqrt{[(k/P_a v_a)(2/k+1)^{(k+1)/(k-1)}]}$$

If G_{350} and G_{192} are the rates of discharge at 350 and $192\ kN/m^2$ respectively, then:

$$G_{350}/G_{192} = 350/192 = 1 \cdot 82$$

or

$$G_{192} = 0 \cdot 55 G_{350}$$

For pressures below $192\ kN/m^2$, use is made of equation 5.24:

$$G = \frac{C_D A_0}{v_2} \sqrt{2P_1 v_1\left(\frac{k}{k-1}\right)\left[1-\left(\frac{P_2}{P_1}\right)^{(k-1)/k}\right]}$$

As before, substituting for $1/v_1 = P_1/P_a v_a$ and $v_2 = v_a$,

$$G = \frac{C_D A_0}{v_2} \sqrt{2P_a v_a\left(\frac{k}{k-1}\right)\left[1-\left(\frac{P_2}{P_1}\right)^{(k-1)/k}\right]}$$

and $G^2 = (C_D A_0/v_a)^2 2 P_a v_a [k/(k-1)][1-(P_2/P_1)^{(k-1)/k}]$
                               $= (C_D A_0/v_a)^2 2 P_a v_a \times 3.5[1-(P_2/P_1)^{0.286}]$

When $P_1 = 192 \text{ kN/m}^2$, $G_{192} = 0.55 G_{350}$ and P_2, atmospheric pressure, is 101·3 kN/m², then:

$$(0.55 G_{350})^2 = (C_D A_0/v_a)^2 2 P_a v_a \times 3.5[1-(101.3/192)^{0.286}]$$

When the final pressure P_1 is reached, the flowrate is to be given by $0.25 G_{350}$. Hence:

$$(0.25 G_{350})^2 = (C_D A_0/v_a)^2 2 P_a v_a \times 3.5(1-(101.3/P_1)^{0.286})$$

Dividing these last two equations by each other gives:

$$\left(\frac{0.55}{0.25}\right)^2 = \frac{1-(101.3/192)^{0.286}}{1-(101.3/P_1)^{0.286}}$$

from which $P_1 = 102.3 \text{ kN/m}^2$

Problem 5.15

Water discharges from the bottom outlet of an open tank 1·5 m by 1 m in cross-section. The outlet is equivalent to an orifice 40 mm diameter with a coefficient of discharge of 0·6. The water level in the tank is regulated by a float valve on the feed supply which shuts off completely when the height of water above the bottom of the tank is 1 m and which gives a flowrate which is directly proportional to the distance of the water surface below this maximum level. When the depth of water in the tank is 0·5 m the inflow and outflow are directly balanced.

As a result of a short interruption in the supply, the water level in the tank falls to 0·25 m above the bottom but is then restored again. How long will it take the level to rise to 0·45 m above the bottom?

Solution

Equation 5.19 relates the mass flowrate G to the head h for the flow through an orifice when the area of the orifice is small in comparison with the area of the pipe:

$$G = C_D A_0 \rho \sqrt{(2gh)}$$

In this problem, let h be the distance of the water level below the maximum depth of 1 m. Then the head above the orifice is equal to $(1-h)$ and

$$G = C_D A_0 \rho \sqrt{[2g(1-h)]}$$

When the tank contains 0·5 m of water, the flowrate may be calculated as:

$$G = 0.6 \times (\pi/4)(0.04)^2 \times 1000\sqrt{(2 \times 9.81 \times 0.5)}$$
$$= 2.36 \text{ kg/s}$$

The input to the tank is stated to be proportional to h, and when the tank is half full the inflow is equal to the outflow,

i.e.
$$2 \cdot 36 = K \times 0 \cdot 5$$
$$K = 4 \cdot 72 \text{ kg/m s}$$

Thus the inflow $= 4 \cdot 72h$ kg/s
and the outflow $= C_D A_0 \rho \sqrt{2g} \sqrt{(1-h)}$ kg/s.
The net rate of filling $= 4 \cdot 72h - C_D A_0 \rho \sqrt{2g} \sqrt{(1-h)}$

$$= 4 \cdot 72h - 3 \cdot 34 \sqrt{(1-h)}$$

Time to fill the tank = mass of water/rate of filling
$$= 1 \times 1 \cdot 5 \times (0 \cdot 45 - 0 \cdot 25) \times 1000/\text{rate}$$
$$= 300/\text{rate}$$

The time to fill from 0·25 to 0·45 m above the bottom of the tank is then given by:

$$\text{time} = \int_{0 \cdot 55}^{0 \cdot 75} \frac{300\,dh}{4 \cdot 72h - 3 \cdot 34 \sqrt{(1-h)}}$$

This integral is most easily solved by graphical means as shown in Fig. 5(c). From Fig. 5c the area under the curve $= 0 \cdot 233$ and the time $= 300 \times 0 \cdot 233 = \underline{\underline{70 \text{ s}}}$.

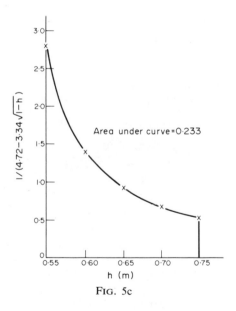

FIG. 5c

Problem 5.16

The flowrate of air at 298 K in a 0·3 m diameter duct is measured with a pitot tube which is used to traverse the cross-section. Readings of the differential pressure recorded on a water manometer are taken with the pitot tube at ten different positions in the cross-section. These positions are so chosen as to be the mid-points of ten concentric annuli each of the same cross-sectional area. The readings are as follows:

Position	1	2	3	4	5
Manometer reading (mm water)	18·5	18·0	17·5	16·8	15·7
Position	6	7	8	9	10
Manometer reading (mm water)	14·7	13·7	12·7	11·4	10·2

The flow is also metered using a 15 cm orifice plate across which the pressure differential is 50 mm on a mercury-under-water manometer. What is the coefficient of discharge of the orifice meter?

Solution

Cross-sectional area of duct $= (\pi/4)(0\cdot3)^2 = 0\cdot0707 \text{ m}^2$.
Area of each concentric annulus $= 0\cdot00707 \text{ m}^2$.
If the diameters of the annuli are designated d_1, d_2, etc.:

$$0\cdot00707 = (\pi/4)(0\cdot3^2 - d_1^2)$$
$$0\cdot00707 = (\pi/4)(d^2 - d_2^2)$$
$$0\cdot00707 = (\pi/4)(d_2^2 - d_3^2), \text{ etc.},$$

and the mid-points of each annulus may be calculated across the duct.
For a pitot tube the velocity may be calculated from the head h as:

$$u = \sqrt{(2gh)}$$

For position 1, $h = 18\cdot5$ mm of water:
The density of the air $= (29/22\cdot4)(273/298) = 1\cdot186 \text{ kg/m}^3$.

$$h = 18\cdot5 \times 10^{-3} \times 1000/1\cdot186$$
$$= 15\cdot6 \text{ m of air}$$

and
$$u = \sqrt{(2 \times 9\cdot81 \times 15\cdot6)}$$
$$= 17\cdot49 \text{ m/s}$$

In the same way, the velocity distribution across the tube may be found, the calculations being shown in the table opposite.
Mass flowrate $G = 1\cdot107 \times 1\cdot186$
$$= 1\cdot313 \text{ kg/s}$$
Considering the orifice, $[1 - (A_0/A_1)^2] = [1 - (0\cdot15/0\cdot3)^2] = 0\cdot938$

$$h = 50 \text{ mmHg under water}$$
$$\equiv 0\cdot05 \times (13\cdot55 - 1) \times 1000/1\cdot186$$
$$= 529 \text{ m of air}$$

and
$$1\cdot313 = C_D(\pi/4)(0\cdot15)^2 \times 1\cdot186\sqrt{(2 \times 9\cdot81 \times 529/0\cdot938)}$$

from which
$$C_D = \underline{\underline{0\cdot61}}$$

Position	Distance from axis of duct (mm)	Manometer reading		Air velocity (u m/s)	Velocity × area of annulus (m³/s)
		(mm water)	(m air)		
1	24	18·5	15·6	17·5	0·124
2	57	18·0	15·17	17·3	0·122
3	75	17·5	14·75	17·0	0·120
4	89	16·8	14·16	16·7	0·118
5	101	15·7	13·23	16·1	0·114
6	111	14·7	12·39	15·6	0·110
7	121	13·7	11·55	·15·1	0·107
8	130	12·7	10·71	14·5	0·103
9	139	11·4	9·61	13·7	0·097
10	147	10·2	8·60	13·0	0·092

Total = 1·107

Although not specifically asked for in the problem, the velocity profile across the duct is plotted in Fig. 5d.

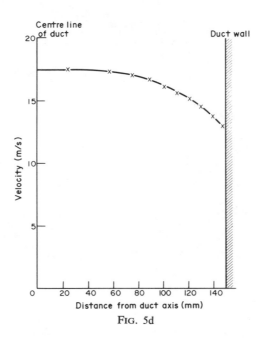

FIG. 5d

Problem 5.17

Explain the principle of operation of the pitot tube and indicate how it can be used in order to measure the total flowrate of fluid in a duct. If a pitot tube is inserted in a circular cross-section pipe in which a fluid is in streamline flow, calculate at what point in the cross-section it should be situated so as to give a direct reading representative of the mean velocity of flow of the fluid.

Solution

The principle of operation of a pitot tube is discussed in Section 5.3.1. It should be emphasised that the pitot tube measures the point velocity of a flowing fluid and not the average velocity so that in order to find the average velocity, a traverse across the duct is necessary. Treatment of typical results is illustrated in Problem 5.16. The point velocity is given by $u = \sqrt{(2gh)}$ where h is the difference of head expressed in terms of the fluid flowing.

For streamline flow the velocity distribution is discussed in Section 3.5.2 where it is shown in equation 3.33 that

$$u_s / u_{max} = 1 - (s^2/r^2)$$

where u_s and u_{max} are the point velocities at a distance s from the wall and at the axis respectively and r is the radius of the pipe. Equation 3.36 relates the average velocity u_{av} to the maximum velocity u_{max} as

$$u_{av} = u_{max}/2$$

Hence when $u_s = u_{av} = u_{max}/2$

$$u_s / u_{max} = (u_{max}/2)/u_{max} = 1 - (s^2/r^2)$$

and
$$0.5 = s^2/r^2$$

and
$$s = 0.707r$$

SECTION 6

PUMPING OF FLUIDS

Problem 6.1

A three-stage compressor is required to compress air from 140 kN/m^2 and 283 K to 4000 kN/m^2. Calculate the ideal intermediate pressures, the work required per kilogram of gas, and the isothermal efficiency of the process. Assume the compression to be adiabatic and the interstage cooling to cool the air to the initial temperature. Show qualitatively, by means of temperature–entropy diagrams, the effect of unequal work distribution and imperfect intercooling, on the performance of the compressor.

Solution

It is shown in Section 6.6.2 that the work done is a minimum when the intermediate pressures P_{i1} and P_{i2} are related to the initial and final pressures P_1 and P_2 by equation 6.50, i.e.

$$P_{i1}/P_1 = P_{i2}/P_{i1} = P_2/P_{i2}$$

In this problem, $P_1 = 140 \text{ kN/m}^2$ and $P_2 = 4000 \text{ kN/m}^2$.

$$\therefore \qquad P_2/P_1 = 28\cdot57$$

and

$$P_{i2}/P_{i1} = P_2/P_{i2} = \sqrt[3]{28\cdot57} = 3\cdot057$$

and

$$P_{i1} = 428 \text{ kN/m}^2$$

$$P_{i2} = 1308 \text{ kN/m}^2$$

The specific volume of the air at the inlet is given by:

$$v_1 = (22\cdot4/29)(283/273)(101\cdot3/140)$$
$$= 0\cdot579 \text{ m}^3/\text{kg}$$

Hence for 1 kg of air, equation 6.51 gives the minimum work of compression in a compressor of n stages as:

$$W = nP_1v_1\frac{\gamma}{\gamma-1}\left[\left(\frac{P_2}{P_1}\right)^{(\gamma-1)/n\gamma} - 1\right]$$
$$W = 3 \times 140{,}000 \times 0\cdot579(1\cdot4/0\cdot4)[(28\cdot57)^{0\cdot4/3\times1\cdot4} - 1]$$
$$= 319{,}170 \text{ J/kg}$$

The isothermal work of compression is given by equation 6.41 as:

$$W_{iso} = P_1 V_1 \ln(P_2/P_1)$$
$$= 140,000 \times 0.579 \ln 28.57$$
$$= 271,740 \text{ J/kg}$$

The isothermal efficiency $= 100 \times 271,740/319,170$
$$= \underline{\underline{85.1\%}}$$

Compression cycles are shown in Figs. 6a and 6b. The former indicates the effect

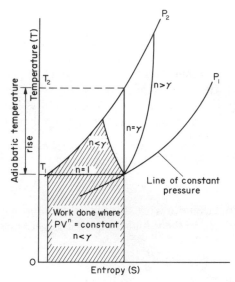

FIG. 6a

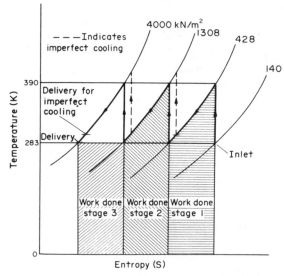

FIG. 6b

of various values of n in PV^n = constant and it is seen that the work done is the area under the temperature–entropy curve. Figure 6b illustrates the three-stage compressor of this problem. The final temperature T_2 is found from $T_2/T_1 = (P_2/P_1)^{(\gamma-1)/\gamma}$ and $T_2 = 390$ K. The dotted lines illustrate the effect of imperfect interstage cooling.

Problem 6.2

A twin-cylinder, single-acting compressor, working at 5 Hz, delivers air at 515 kN/m² pressure at the rate of 0.2 m³/s. If the diameter of the cylinder is 20 cm, the cylinder clearance ratio 5%, and the temperature of the inlet air 283 K, calculate the length of stroke of the piston and the delivery temperature.

Solution

For adiabatic conditions, PV^γ = constant

and
$$P_2/P_1 = (T_2/T_1)^{\gamma/(\gamma-1)}$$
or
$$T_2 = T_1(P_2/P_1)^{(\gamma-1)/\gamma}$$

Thus the delivery temperature $= 283(515/101.3)^{0.4/1.4}$
$$= \underline{\underline{500 \text{ K}}}$$

The volume handled per cylinder $= 0.2/2 = 0.1$ m³/s.
Volume per stroke $= 0.1/5 = 0.02$ m³/s at 515 kN/m².
Volume at inlet condition $= 0.02 \times 283/500 = 0.0126$ m³/s.
From equation 6.47, $0.0126 = V_s[1 + c - c(P_2/P_1)^{1/\gamma}]$ where c is the clearance and V_s the swept volume.

∴
$$0.0126 = V_s[1 + 0.05 - 0.05(515/101.3)^{1/1.4}]$$
and
$$V_s = 0.0142 \text{ m}^3$$

∴
$$(\pi/4)(0.2)^2 \times \text{stroke} = 0.0142$$
and
$$\text{the stroke} = \underline{\underline{0.45 \text{ m}}}$$

Problem 6.3

A single-stage double-acting compressor running at 3 Hz is used to compress air from 110 kN/m² and 282 K to 1150 kN/m². If the internal diameter of the cylinder is 20 cm, the length of stroke 25 cm, and the piston clearance 5%, calculate:
(a) the maximum capacity of the machine, referred to air at the initial temperature and pressure; and
(b) the theoretical power requirements under isentropic conditions.

Solution

The volume per stroke $= 2 \times (\pi/4)(0 \cdot 2)^2 \times 0 \cdot 25$
$$= 0 \cdot 0157 \text{ m}^3$$
The compression ratio $= 1150/110 = 10 \cdot 45$.
The swept volume V_s is given by equation 6·47:

$$0 \cdot 0157 = V_s[1 + 0 \cdot 05 - 0 \cdot 05(10 \cdot 45)^{1/1 \cdot 4}]$$
and $$V_s = 0 \cdot 0217 \text{ m}^3$$

From equation 6.46 the work of compression/cycle is given by:

$$W = P_1(V_1 - V_4)(\gamma/(\gamma - 1))[(P_2/P_1)^{(\gamma-1)/\gamma} - 1]$$

and from equation 6.48, substituting for $(V_1 - V_4)$, gives:

$$W = P_1 V_s[1 + c - c(P_2/P_1)^{1/\gamma}][\gamma/(\gamma - 1)][(P_2/P_1)^{(\gamma-1)/\gamma} - 1]$$
$$= 110,000 \times 0 \cdot 0157(1 \cdot 4/0 \cdot 4)[(10 \cdot 45)^{0 \cdot 286} - 1]$$
$$= 5781 \text{ J}$$

The theoretical power requirement $= 3 \times 5781 = 17,340 \text{ W}$
$$= \underline{\underline{17 \cdot 3 \text{ kW}}}$$

The capacity $= 3 \times 0 \cdot 0157 = \underline{\underline{0 \cdot 047 \text{ m}^3/\text{s}}}$

Problem 6.4

Methane is to be compressed from atmospheric pressure to 30 MN/m² in four stages.

Calculate the ideal intermediate pressures and the work required per kilogram of gas. Assume compression to be isentropic and the gas to behave as an ideal gas. Indicate on a temperature–entropy diagram the effect of imperfect intercooling on the work done at each stage.

Solution

The ideal intermediate pressures correspond to the situation when the compression ratios in each stage are equal. If the initial, intermediate, and final pressures from this compressor are designated P_1, P_2, P_3, P_4, and P_5, then:

$$P_2/P_1 = P_3/P_2 = P_4/P_3 = P_5/P_4 = P_5/P_1$$

as in problem 6.1

$$P_5/P_1 = (30,000/101 \cdot 3) = 296 \cdot 2$$
and $$(P_5/P_1)^{0 \cdot 25} = 4 \cdot 148$$

Hence $$P_2 = 4 \cdot 148 P_1 = 4 \cdot 148 \times 101 \cdot 3 = \underline{\underline{420 \text{ kN/m}^2}}$$

$$P_3 = 4 \cdot 148 P_2 = \underline{\underline{1 \cdot 74 \text{ MN/m}^2}}$$

$$P_4 = 4 \cdot 148 P_3 = \underline{\underline{7 \cdot 23 \text{ MN/m}^2}}$$

The work required per kilogram of gas is given by equation 6.51:

$$W = n P_1 V_1 \frac{\gamma}{\gamma - 1} \left[\left(\frac{P_5}{P_1} \right)^{(\gamma - 1)/n\gamma} - 1 \right]$$

For methane, the molecular weight = 16 kg/kmol and the specific volume at STP = $22 \cdot 4/16 = 1 \cdot 40 \text{ m}^3/\text{kg}$.

If $\gamma = 1 \cdot 4$, the work per kilogram is given by:

$$W = 4 \times 101{,}300 \times 1 \cdot 40 (1 \cdot 4/0 \cdot 4)[(296 \cdot 2)^{0 \cdot 4/4 \times 1 \cdot 4} - 1]$$

$$= 710{,}940 \text{ J/kg}$$

$$\equiv \underline{\underline{711 \text{ kJ/kg}}}$$

The effect of imperfect cooling is shown in Figs. 6a and 6b.

Problem 6.5

An air-lift raises $0 \cdot 01 \text{ m}^3/\text{s}$ of water from a well 100 m deep through a 100 mm diameter pipe. The level of the water is 40 m below the surface. The air consumed is $0 \cdot 1 \text{ m}^3/\text{s}$ of free air compressed to 800 kN/m^2.

Calculate the efficiency of the pump and the mean velocity of the mixture in the pipe.

Solution

Mass flow of water = $0 \cdot 01 \times 1000 = 10 \text{ kg/s}$.

Work done = $10 \times 40 \times 981 = 3924 \text{ W}$.

Volume of air used = $0 \cdot 1 \text{ m}^3/\text{s}$.

Equation 6.42 enables the work needed to compress $0 \cdot 1 \text{ m}^3$ of air to be calculated as:

$$P_1 V_1 [\gamma/(\gamma - 1)][(P_2/P_1)^{(\gamma - 1)/\gamma} - 1]$$

$$= 101{,}300 \times 0 \cdot 1 \times 1 \cdot 4/0 \cdot 4[(800/101 \cdot 3)^{0 \cdot 286} - 1]$$

$$= 28{,}750 \text{ J}$$

Power required for this compression = $28{,}750 \text{ J/s}$

$$\equiv 28{,}750 \text{ W}$$

Efficiency = $3924 \times 100/28{,}750$

$$= \underline{\underline{13 \cdot 7\%}}$$

The mean velocity depends upon the pressure of air in the pipe. If the air pressure at the bottom of the well is taken as being 60 m of water, i.e. $60 \times 1000 \times 9 \cdot 81 \times 10^{-3} = 588 \cdot 6 \text{ kN/m}^2$, and the pressure at the surface is atmospheric, the mean pressure is:

$$(101 \cdot 3 + 588 \cdot 6)/2 = 345 \text{ kN/m}^2$$

The specific volume of air at this pressure (temperature taken as 273 K) = $(22\cdot4/29)(101\cdot3/345) = 0\cdot227$ m^3/kg.

Specific volume at STP = $22\cdot4/29 = 0\cdot772$ m^3/kg.

∴ mass of air = $0\cdot1/0\cdot772 = 0\cdot13$ kg/s.

Mean volume of air = $0\cdot13 \times 0\cdot227 = 0\cdot0295$ m^3/s.

Volume of water = $0\cdot01$ m^3/s.

Total volumetric flow = $0\cdot0395$ m^3/s.

Area of pipe = $(\pi/4)0\cdot1^2 = 0\cdot00785$ m^2.

∴   mean velocity = $0\cdot0395/0\cdot00785$

                                                  $= 5\cdot03$ m/s

Problem 6.6

In a single-stage compressor:

Suction pressure = $101\cdot3$ kN/m^2.
Suction temperature = 283 K.
Final pressure = 380 kN/m^2.
Compression is adiabatic.

If each new charge is heated 18 K by contact with the clearance gases, calculate the maximum temperature attained in the cylinder.

Solution

The compression ratio = $380/101\cdot3 = 3\cdot75$.

On the first stroke, the air enters at 283 K and is compressed adiabatically to 380 kN/m^2. The exit temperature T_2 is then given by:

$$T_2/T_1 = (P_2/P_1)^{(\gamma-1)/\gamma}$$
$$= 3\cdot75^{0\cdot286} = 1\cdot459$$

Hence $T_2 = 1\cdot459 \times 283 = 413$ K

Thus the clearance volume gases which remain in the cylinder are able to raise the next cylinder full of air by 18 K leaving the cylinder and its contents at $283 + 18 = 301$ K. After compression, the exit temperature is given by

$$T = 301 \times 3\cdot75^{0\cdot286}$$
$$= 439\cdot2 \text{ K}$$

On each subsequent stroke the inlet temperature is always 301 K and hence the maximum temperature attained is $439\cdot2$ K.

Problem 6.7

A single-acting reciprocating pump has a cylinder diameter of 115 mm and a stroke of 230 mm. The suction line is 6 m long and 50 mm diameter and the level of the water in the suction tank is 3 m below the cylinder of the pump. What is the maximum speed at which the pump can run without an air vessel if separation is not to occur in the suction line? The piston undergoes approximately simple harmonic motion. Atmospheric pressure is equivalent to a head of 10·4 m of water and separation occurs at pressure corresponding to a head of 1·22 m of water.

Solution

The tendency for separation to occur will be greatest at the inlet to the cylinder and at the beginning of the suction stroke.

If the maximum speed of the pump is N Hz, the angular velocity of the driving mechanism is $2\pi N$ radians/s.

The acceleration of the piston $= 0·5 \times 0·23(\pi N)^2 \cos(2\pi N)$ m/s^2.

The maximum acceleration, when $t = 0$, $= 4·54 N^2$ m/s^2.

Maximum acceleration of the liquid in the suction pipe is given by:

$$(0·115/0·05)^2 \times 4·54N^2 = 24·02 \, N^2 \quad \text{m/s}$$

Accelerating force on the liquid $= 24·02N^2(\pi/2)(0·05)^2 \times 6 \times 1000$.

Pressure drop in suction line due to acceleration $= 24·02N^2 \times 6 \times 1000$
$$= 1·44 \times 10^5 N^2 \text{ N/m}^2$$
$$= 1·44 \times 10^5 \, N^2/1000 \times 9·81$$
$$= 14·69N^2 \quad \text{m of water}$$

Pressure head at the cylinder when separation is about to occur,

$$1·22 = 10·4 - 3·0 - 14·69N^2 \text{ m of water}$$

and
$$\underline{\underline{N = 0·65 \text{ Hz}}}$$

Problem 6.8

An air-lift pump is used for raising 800 cm^3/s of a liquid of specific gravity 1·2 to a height of 20 m. Air is available at 450 kN/m^2. If the efficiency of the pump is 30%, calculate the power requirement, assuming isentropic compression of the air ($\gamma = 1·4$).

Solution

Mass of flowrate of liquid $= 800 \times 10^{-6} \times 1·2 \times 1000$
$$= 0·96 \text{ kg/s}$$

Work done $= 0·96 \times 20 \times 9·81$
$$= 188·4 \text{ W}$$

Actual work of expansion of air $= 188·4/0·3 = 627·8$ W.

The mass of air required/s is given by equation 6.26 as:

$$W = P_a v_a m \ln(P/P_a)$$
$$= P_a V_a \ln(P/P_a)$$

where V_a is the volume of air at STP

i.e. $627 \cdot 8 = 101,300 \, V_a \, \ln(450/101 \cdot 3)$

and $V_a = 0 \cdot 0042 \text{ m}^3$

The work done in the isentropic compression of this air is given by equation 6.42:

$$\text{Work} = P_1 V_1 [\gamma/(\gamma - 1)][(P_2/P_1)^{(\gamma - 1)/\gamma} - 1]$$
$$= 101,300 \times 0 \cdot 0042 (1 \cdot 4/0 \cdot 4)[(450/101 \cdot 3)^{0 \cdot 286} - 1]$$
$$= 792 \text{ J}$$

Power required $= 792 \text{ J/s} \equiv 792 \text{ W} \equiv \underline{0 \cdot 79 \text{ kW}}$.

Problem 6.9

A single-acting air compressor supplies $0 \cdot 1 \text{ m}^3/\text{s}$ of air (at STP) compressed to 380 kN/m^2 from $101 \cdot 3 \text{ kN/m}^2$ pressure. If the suction temperature is $288 \cdot 5 \text{ K}$, the stroke is 250 mm, and the speed is 4 Hz, find the cylinder diameter. Assume the cylinder clearance is 4% and compression and re-expansion are isentropic ($\gamma = 1 \cdot 4$). What is the theoretical power required for the compression?

Solution

The compression ratio $P_2/P_1 = 380/101 \cdot 3 = 3 \cdot 75$.
Volume per stroke $= (0 \cdot 1/4 \cdot 0)(289/273)$
$$= 0 \cdot 0265 \text{ m}^3$$
The swept volume may be calculated from equation 6.47:

$$\text{swept volume} = V_1 - V_4 = V_s [1 + c - c \, (P_2/P_1)^{1/\gamma}]$$

where $c = $ clearance.

Hence $0 \cdot 0265 = V_s [1 + 0 \cdot 04 - 0 \cdot 04(3 \cdot 75)^{0 \cdot 714}]$

and $V_s = 0 \cdot 0283 \text{ m}^3$

Cross-sectional area of cylinder $=$ volume/stroke
$$= 0 \cdot 0283/0 \cdot 25 = 0 \cdot 113 \text{ m}^2$$

Hence the cylinder diameter $= \sqrt{0 \cdot 113/(\pi/4)} = \underline{0 \cdot 38 \text{ m}}$.

The work of compression per cycle is given by equation 6.46:

$$\text{Work} = P_1(V_1 - V_4)[\gamma/(\gamma - 1)][(P_2/P_1)^{(\gamma - 1)/\gamma} - 1]$$
$$= 101,300 \times 0 \cdot 0265(1 \cdot 4/0 \cdot 4)[(3 \cdot 75)^{0 \cdot 286} - 1]$$
$$= 4278 \text{ J/cycle}$$

Total theoretical power $= 4 \times 4278 = 17,110 \text{ W}$
$$\equiv \underline{17 \cdot 1 \text{ kW}}$$

Problem 6.10

Air at 290 K is compressed from 101·3 to 2000 kN/m² pressure in a two-stage compressor operating with a mechanical efficiency of 85%. The relation between pressure and volume during the compression stroke and expansion of the clearance gas is $PV^{1·25} = $ constant. The compression ratio in each of the two cylinders is the same and the interstage cooler may be taken as perfectly efficient. If the clearances in the two cylinders are 4% and 5% respectively, calculate:

(a) the work of compression per unit mass of gas compressed,
(b) the isothermal efficiency,
(c) the isentropic efficiency ($\gamma = 1·4$),
(d) the ratio of the swept volumes in the two cylinders.

Solution

(a) The overall compression ratio $= 2000/101·3 = 19·74$.
  Specific volume of air at 290 K $= (22·4/29)(290/273) = 0·821$ m³/kg.
  Equation 6.51 gives the minimum work of compression for an *n*-stage compressor as:

$$W = nP_1V_1\frac{\gamma}{\gamma-1}\left[\left(\frac{P_2}{P_1}\right)^{(\gamma-1)/n\gamma} - 1\right]$$

$$= 2 \times 101{,}300 \times 0·821 \times (1·25/0·25)[(19·74)^{0·25/2·5} - 1]$$

$$= 289{,}000 \text{ J/kg} \equiv 289 \text{ kJ/kg}$$

  The actual work of compression $= 289/0·85 = \underline{340 \text{ kJ/kg}}$.

(b) The work done in the isothermal compression of 1 kg of gas is given by equation 6.41:

$$W = P_1V_1\ln(P_2/P_1)$$

$$= 101{,}300 \times 0·821 \ln 19·74$$

$$= 248{,}000 \text{ J/kg} = 248 \text{ kJ/kg}$$

  The isothermal efficiency $= 100 \times 248/340 = \underline{72·9\%}$.

(c) The work done in the isentropic compression of 1 kg of gas is given by equation 6.42:

$$W = P_1V_1\frac{\gamma}{\gamma-1}\left[\left(\frac{P_2}{P_1}\right)^{(\gamma-1)/\gamma} - 1\right]$$

$$= 101{,}300 \times 0·821 \times (1·4/0·4)((19·74)^{0·4/1·4} - 1)$$

$$= 329{,}000 \text{ J/kg} \equiv 392 \text{ kJ/kg}$$

  The isentropic efficiency $= 100 \times 392/340 = \underline{115\%}$.

(d) From equation 6.52, the volume swept out in the first cylinder in compressing 1 kg of gas is given by:

$$\text{volume admitted} = V_{s1}[1 + c_1 - c_1(P_2/P_1)^{1/n\gamma}]$$

As the same mass of gas passes through each of the cylinders, and if the interstage coolers are 100% efficient, the ratio of the volumes admitted to successive cylinders is $(P_1/P_2)^{1/n}$. The volume of gas admitted to the second cylinder is then:

$$= V_{s2}[1 + c_2 - c_2(P_2/P_1)^{1/n\gamma}] = V_{s1}[1 + c_1 - c_1(P_2/P_1)^{1/n\gamma}](P_1/P_2)^{1/n}$$

and

$$\frac{V_{s1}}{V_{s2}} = \frac{1 + c_2 - c_2(P_2/P_1)^{1/n\gamma}}{1 + c_1 - c_1(P_2/P_1)^{1/n\gamma}} \left(\frac{P_2}{P_1}\right)^{1/n}$$

$$= \frac{1 + 0.05 - 0.05(19.74)^{1/2 \cdot 5}}{1 + 0.04 - 0.04(19.74)^{1/2 \cdot 5}} (19.74)^{1/2}$$

$$= \underline{\underline{4.14}}$$

Problem 6.11

Explain briefly the significance of the "specific speed" of a centrifugal or axial-flow pump.

A pump is designed to be driven at 10 Hz and to operate at a maximum efficiency when delivering 0.4 m³/s of water against a head of 20 m. Calculate the specific speed. What type of pump does this value suggest? A pump built for these operating conditions has a measured overall efficiency of 70%. The same pump is now required to deliver water at 30 m head. At what speed should the pump be driven if it is to operate at maximum efficiency? What will be the new rate of delivery and the power required?

Solution

Specific speed is discussed in Section 6.2.3, where it is shown to be equal to $N_s = NQ^{1/2}/(gh)^{3/4}$. This expression is dimensionless providing that the pump speed, throughput, and head are expressed in consistent units.

In this problem, $N = 10$ Hz, $Q = 0.4$ m³/s, and $h = 20$ m.

$$\therefore \qquad N_s = 10 \times (0.4)^{0.5}/(9.81 \times 20)^{0.75}$$

$$= \underline{\underline{0.121}}$$

Reference should be made to specialist texts on pumps where classifications of pump types as a function of specific speed are presented. This example would suggest a centrifugal pump.

From equation 6.17, $Q \propto N$ and

$$Q_1/Q_2 = N_1/N_2$$

and from equations 6.15 and 6.16, $h \propto N^2$ and

$$h_1/h_2 = (N_1/N_2)^2$$

$$\therefore \qquad 20/30 = (10/N_2)^2$$

and

$$N_2 = \underline{\underline{12.24 \text{ Hz}}}$$

From the above, $0.4/Q_2 = 10/12.24$

and $Q_2 = 0.49 \text{ m}^3/\text{s}$

Power required $= (1/n)(\text{mass flow} \times \text{head} \times g)$
$$= (1/0.7)(0.49 \times 1000 \times 30 \times 9.81)$$
$$= 206 \text{ W}$$

Problem 6.12

A centrifugal pump is to be used to extract water from a condenser in which the vacuum is 640 mm of mercury. At the rated discharge, the net positive suction head must be at least 3 m above the cavitation vapour pressure of 710 mm mercury vacuum. If losses in the suction pipe account for a head of 1·5 m, what must be the least height of the liquid level in the condenser above the pump inlet?

Solution

The system is illustrated in Fig. 6c. From an energy balance, the head at the suction point of the pump h_i may be found as:

$$h_i = (P_0/\rho h) + x - (u_i^2/2g) - h_f$$

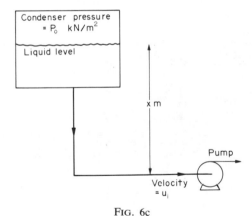

FIG. 6c

In this problem, the losses in the suction pipe $= 1.5$ m, i.e.

$$(u_i^2/2g) + h_f = 1.5$$

The net positive suction head (NPSH) is discussed in Section 6.2.3 where it is shown that:

$$\text{NPSH} = h_i - (P_v/\rho g)$$

where P_v is the vapour pressure of the liquid being pumped. Use of this equation will give the minimum height x as:

$$3 = (P_v/\rho g) + x - 1\cdot5 - (P_v/\rho g)$$
$$P_0 = 760 - 640 = 120 \text{ mmHg} = 16,000 \text{ N/m}^2$$
$$P_v = 760 - 710 = 50 \text{ mmHg} = 6670 \text{ N/m}^2$$
$$\rho = 1000 \text{ kg/m}^3, \quad g = 9\cdot81 \text{ m/s}^2$$

$$\therefore \qquad x = 3 + 1\cdot5 - (16{,}000 + 6670)/(1000 \times 9\cdot81)$$
$$= \underline{\underline{3\cdot55 \text{ m}}}$$

Problem 6.13

What is meant by the Net Positive Suction Head (NPSH) required by a pump? Explain why it exists and how it can be made as low as possible. What happens if the necessary NPSH is not provided?

A centrifugal pump is to be used to circulate liquid (sp. gr. 0·80 and viscosity 0·5 mN s/m²) from the reboiler of a distillation column through a vaporiser at the rate of 400 cm³/s, and to introduce the superheated liquid above the vapour space in the reboiler which contains liquid to a depth of 0·7 m. Suggest a suitable layout if a smooth-bore 25 mm pipe is to be used. The pressure of the vapour in the reboiler is 1 kN/m² and the NPSH required by the pump is 2 m of liquid (Fig. 6d).

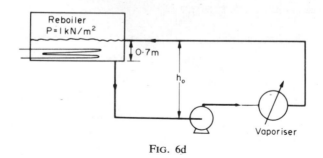

FIG. 6d

Solution

The topic of net positive suction head is discussed in Section 6.2.3. The NPSH is the amount by which the pressure at the suction point of the pump, expressed as head of liquid, must exceed the vapour pressure of the liquid. It is equal to (from equation 6.25):

$$\text{NPSH} = (P_0 - P_v)/\rho g + h_0 - h_f$$

where P_0 and P_v are the pressure acting on the liquid being pumped and the vapour pressure respectively, $(h_f + u_i^2/2g)$ the head loss in the suction piping system, and h_0 the liquid level above the pump inlet.

The NPSH varies with pump throughput, and pump manufacturers publish curves relating this value to capacity and speed of each pump. If a pump is subjected to a large suction lift and/or to heavy friction losses, the pressure of the liquid at the "eye" of the impeller may fall below its vapour pressure at the temperature concerned. Cavitation will then occur when the liquid vaporises and the pump then has to handle a mixture of liquid, vapour, and air. At this point, the capacity of the pump is severely reduced and mechanical damage may occur. In order to solve this problem, the head loss in the pipe and the kinetic energy losses must be found on the suction side of the pump.

Volumetric flowrate $= 400 \times 10^{-6} \, \text{m}^3/\text{s}$.
Cross-sectional area of pipe $= (\pi/4)(0{\cdot}025)^2 = 0{\cdot}00049 \, \text{m}^2$.
Velocity in the pipe, $u = 400 \times 10^{-6}/0{\cdot}00049 = 0{\cdot}816 \, \text{m/s}$.
Reynolds number $= \rho u d/\mu$
$$= (1000 \times 0{\cdot}8) \times 0{\cdot}816 \times 0{\cdot}025/0{\cdot}5 \times 10^{-3}$$
$$= 3{\cdot}27 \times 10^4$$

For a smooth pipe, the friction factor $R/\rho u^2$ is found from Fig. 3.7 to equal 0·0028.

From equation 3.17 the head loss due to friction h_f is given by:

$$h_f = 8(R/\rho u^2)(l/d)(u^2/2g)$$

$\therefore \qquad h_f/l = 8 \times 0{\cdot}0028 \times 0{\cdot}025(0{\cdot}816^2/2 \times 9{\cdot}81)$

$$= 0{\cdot}0304 \, \text{m/m of pipe}$$

As the liquid being pumped is at its boiling point, $(P_0 - P_v)/\rho g = 0$ and the NPSH equation becomes:

$$2 = h_0 - h_f l$$
$$h_0 = 2{\cdot}0 + h_f l$$

Thus the height h_0 between the liquid level and the pump depends upon the length of pipe between the reboiler and the pump. If this length is 10 m, the minimum value of h_0 is 2·3 m.

SECTION 7

HEAT TRANSFER

Problem 7.1

Calculate the time taken for the distant face of a brick wall, of thermal diffusivity, $D_H = 0.0042$ cm^2/s and thickness $l = 0.45$ m, initially at 290 K, to rise to 470 K if the near face is suddenly raised to a temperature of $\theta' = 870$ K and maintained at that temperature. Assume that all the heat flow is perpendicular to the faces of the wall and that the distant face is perfectly insulated.

Solution

The temperature at any distance x from the near face at time t is given by Equation 7.33:

$$\theta = \sum_{N=0}^{N=\infty} (-1)^N \theta'\{\text{erfc}[(2lN + x)/(2\sqrt{D_H t})] + \text{erfc}[2(N + 1)l - x/(2\sqrt{D_H t})]\}$$

and the temperature at the distant face is:

$$\theta = \sum_{N=0}^{N=\infty} (-1)^N \theta'\{2\,\text{erfc}[(2N + 1)l]/(2\sqrt{D_H t})\}$$

Choosing the temperature scale such that the initial temperature is everywhere zero,

$$\theta/2\theta' = (470 - 290)/2(870 - 290) = 0.155$$

$$D_H = 0.0042 \text{ cm}^2/\text{s} \quad \text{or} \quad 4.2 \times 10^{-7} \text{ m}^2/\text{s} \quad \text{and} \quad \sqrt{D_H} = 6.481 \times 10^4$$

$$l = 0.45 \text{ m}$$

Thus
$$0.155 = \sum_{N=0}^{N=\infty} (-1)^N \text{ erfc } 347(2N + 1)/t^{0.5}$$

$$= \text{erfc}(347t^{-0.5}) - \text{erfc}(1042t^{-0.5}) + \text{erfc}(1736t^{-0.5})$$

Considering the first term only,

$$347t^{-0.5} = 1.0$$

and
$$t = 1.204 \times 10^5 \text{ s}$$

On inspection, the second and higher terms are negligible compared with the first term at this value of t

and hence
$$t = 0.120 \text{ Ms } (33.5 \text{ hr})$$

Problem 7.2

Calculate the time for the distant face to reach 470 K under the same conditions except that the distant face is not perfectly lagged. Instead, a very large thickness of material of the same thermal properties as the brickwork is stacked against it.

Note: $p^{-1} e^{-k\sqrt{p}}$ is the Laplace transform of erfc $k/(2\sqrt{t})$.

Tables for erfc x for various values of x are given on p. 373 of *Conduction of Heat in Solids* by Carslaw and Jaeger.

Solution

In this case, the problem involves conduction of heat in an infinite medium in which it is required to determine the time at which a point 0·45 m from the heated face reaches 470 K.

Thus the boundary conditions are:

$$\theta = 0, \quad t = 0; \quad \theta = \theta', \quad t > 0 \quad \text{for all values of } x$$
$$\theta = (870 - 290) = 580 \text{ K}, \quad x = 0, \quad t > 0$$
$$\theta = 0, \quad x = \infty, \quad t > 0$$
$$\theta = 0, \quad x = 0, \quad t = 0$$

From equation 7.24:

$$\frac{\partial \theta}{\partial t} = D_H \left(\frac{\partial^2 \theta}{\partial x^2} + \frac{\partial^2 \theta}{\partial y^2} + \frac{\partial^2 \theta}{\partial z^2} \right) = D_H \frac{\partial^2 \theta}{\partial x^2} \quad \text{in this case.}$$

The Laplace transform of

$$\theta = \bar{\theta} = \int_0^\infty \theta \, e^{-pt} \, dt \tag{i}$$

and hence

$$\frac{\partial^2 \bar{\theta}}{\partial t^2} = \frac{p}{D_H} \bar{\theta} - \frac{\theta_{t=0}}{D_H} \tag{ii}$$

Integrating (ii)

$$\bar{\theta} = B_1 e^{x\sqrt{(p/D_H)}} + B_2 e^{-x\sqrt{(p/D_H)}} + \theta_{t=0}/p \tag{iii}$$

and

$$\frac{\partial \bar{\theta}}{\partial x} = B_1 \sqrt{(p/D_H)} \, e^{x\sqrt{(p/D_H)}} - B_2 \sqrt{(p/D_H)} \, e^{-x\sqrt{(p/D_H)}} \tag{iv}$$

In this case,

$$\bar{\theta}_{\substack{x=0 \\ t>0}} = \int_0^\infty \theta' \, e^{-pt} \, dt = \theta'/p$$

and

$$\left(\frac{\overline{\partial \theta}}{\partial t} \right)_{\substack{x=0 \\ t \to 0}} = \int_0^\infty \left(\frac{\partial \theta}{\partial t} \right) e^{-pt} \, dt = 0$$

Substituting the boundary conditions in equations (iii) and (iv):

$$\bar{\theta}_{\substack{x=0 \\ t>0}} = \theta'_{\substack{x=0 \\ t>0}}/p = B_1 + B_2 + \theta_{t=0}/p$$

or

$$B_1 + B_2 = \theta'_{\substack{x=0 \\ t>0}}/p$$

and $\qquad\qquad \left(\dfrac{\partial \bar{\theta}}{\partial t}\right)_{\substack{t>0 \\ x=\infty}} = 0 = B_1\sqrt{(p/D_H)}\,e^{\infty} - B_2\sqrt{(p/D_H)}\,e^{-\infty}$

$\therefore \qquad\qquad B_1\sqrt{(p/D_H)} = 0 \quad \text{and} \quad B_1 = 0, \quad B_2 = \theta'_{\substack{x=0 \\ t>0}}/p$

From (iii), $\qquad\qquad \bar{\theta} = B_2\,e^{-x\sqrt{(p/D_H)}} = \theta'p^{-1}\,e^{-k\sqrt{p}} \quad \text{where } k = x/\sqrt{D_H}$

The Laplace transform of
$$p^{-1}\,e^{-k\sqrt{p}} = \text{erfc}\,k/2\sqrt{t}$$
and $\qquad\qquad \theta = \theta'_{\substack{t>0 \\ x=0}}\,\text{erfc}[(x/\sqrt{D_H})(1/2\sqrt{t})] \qquad\qquad\qquad\qquad (v)$

When $x = 0.45$ m, $\theta = (470 - 290) = 180$ K, and hence in (v), with $D_H = 4.2 \times 10^{-7}$ m^2/s,
$$(180/580) = \text{erfc}[(0.45/6.481 \times 10^{-4})(1/2\sqrt{t})] = 0.31$$
$\therefore \qquad\qquad (0.45/6.481 \times 10^{-4})/2\sqrt{t} = 0.73$

and $\qquad\qquad\qquad t = 2.26 \times 10^5$ s

or $\qquad\qquad\qquad \underline{\underline{t = 0.226 \text{ Ms } (62.8 \text{ h})}}$

As an alternative method of solution, Schmidt's method will be used with the construction shown in Figure 7a. In this case $\Delta x = 0.1$ m and it is seen that at $x = 0.45$ m the temperature is 470 K after a time $20\Delta t$.

From equation 7.42:
$$\Delta t = (0.1)^2/(2 \times 4.2 \times 10^{-7}) = 1.191 \times 10^4 \text{ s}$$

and hence the required time,
$$t = 20 \times 1.191 \times 10^4 = 2.38 \times 10^5 \text{ s}$$
or $\qquad\qquad\qquad \underline{\underline{0.238 \text{ Ms } (66.1 \text{ h})}}$

The variation is due to inaccuracies introduced by choosing coarse increments of x.

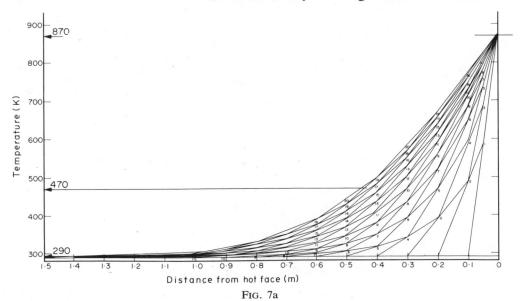

FIG. 7a

Problem 7.3

Benzene vapour, at atmospheric pressure, condenses on a plane surface 2 m long and 1 m wide maintained at 300 K and inclined at an angle of 45° to the horizontal. Plot the thickness of the condensate film and the point heat transfer coefficient against distance from the top of the surface.

Solution

At 101·3 kN/m², benzene condenses at $T_s = 353$ K. With a wall temperature of $T_w = 300$ K, the film properties may be taken at a mean temperature of 327 K:

$$\mu = 4·3 \times 10^{-4} \text{ N s/m}^2$$
$$\rho = 860 \text{ kg/m}^3$$
$$k = 0·151 \text{ W/m K}$$
$$\lambda = 423 \text{ kJ/kg} = 4·23 \times 10^5 \text{ J/kg}$$

Thus in equation 7.100

$$s = \{[4\mu k(T_s - T_w)x]/(g \sin \phi \lambda \rho^2)\}^{0·25}$$
$$= \{[4 \times 4·3 \times 10^{-4} \times 0·151(353 - 300)x]/(9·81 \sin 45 \times 4·23 \times 10^5 \times 860^2)\}^{0·25}$$
$$= 2·82 \times 10^{-4} x^{0·25} \text{ m}$$

Similarly, in equation 7.101:

$$h = \{(\rho^2 g \sin \phi \lambda k^3)/[4\mu(T_s - T_w)x]\}^{0·25}$$
$$= \{(860^2 \times 9·81 \sin 45 \times 4·23 \times 10^5 \times 0·151^3)/[4 \times 4·3 \times 10^{-4}(353 - 300)x]\}^{0·25}$$
$$= 535·1x^{-0·25} \text{ W/m}^2 \text{ K}$$

Values of x between 0 and 2·0 m increasing in increments of 0·20 m are now substituted in these equations with the following results, which are plotted in Figure 7b.

x (m)	$x^{0·25}$	$x^{-0·25}$	s (m)	h (W/m² K)
0	0	∞	0	∞
0·1	0·562	1·778	$1·58 \times 10^{-4}$	951
0·2	0·669	1·495	$1·89 \times 10^{-4}$	800
0·4	0·795	1·258	$2·24 \times 10^{-4}$	673
0·6	0·880	1·136	$2·48 \times 10^{-4}$	608
0·8	0·946	1·057	$2·67 \times 10^{-4}$	566
1·0	1·000	1·000	$2·82 \times 10^{-4}$	535
1·2	1·047	0·956	$2·95 \times 10^{-4}$	512
1·4	1·088	0·919	$3·07 \times 10^{-4}$	492
1·6	1·125	0·889	$3·17 \times 10^{-4}$	476
1·8	1·158	0·863	$3·27 \times 10^{-4}$	462
2·0	1·189	0·841	$3·35 \times 10^{-4}$	450

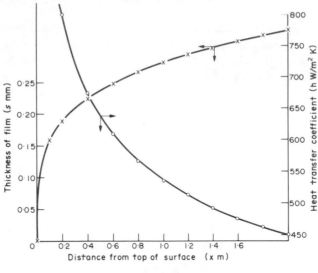

Fig. 7b

Problem 7.4

It is desired to warm 0·9 kg/s of air from 283 to 366 K by passing it through the pipes of a bank consisting of 20 rows with 20 pipes in each row. The arrangement is in-line with centre to centre spacing, in both directions, equal to twice the pipe diameter. Flue gas, entering at 700 K and leaving at 366 K, with a free flow mass velocity of 10 kg/m² s, is passed across the outside of the pipes.

Neglecting gas radiation, how long should the pipes be?

For simplicity, outer or inner pipe diameter may be taken as 12 mm.

Values of k and μ, which may be used for both air and flue gases, are given below. The specific heat of air and flue gases is 1·0 kJ/kg K.

Temperature (K)	Thermal conductivity k(W/m K)	Viscosity μ (mN s/m²)
250	0·022	0·0165
500	0·044	0·0276
800	0·055	0·0367

Solution

Heat load

$$Q = 0·9 \times 1·0(366 - 283) = 74·7 \text{ kW}$$

Temperature driving force

$$\theta_1 = (700 - 366) = 334 \text{ K}, \quad \theta_2 = (366 - 283) = 83 \text{ K}$$

and from equation 7.10:
$$\theta_m = (334 - 83)/\ln(334/83) = 180 \text{ K}$$

Film coefficients

Inside:

In equation 7.50:
$$h_i d_i / k = 0.023 (dG'/\mu)^{0.8} (C_p \mu / k)^{0.4}$$

where $d_i = 12$ mm or 1.2×10^{-2} m.

The mean air temperature $= 0.5(366 + 283) = 325$ K and $k = 0.029$ W/m K.

Cross area of one tube $= (\pi/4)(1.2 \times 10^{-2})^2 = 1.131 \times 10^{-4}$ m^2 and area for flow $= (20 \times 20)1.131 \times 10^{-4} = 4.52 \times 10^{-2}$ m^2.

Thus, mass velocity $G = 0.9/(4.52 \times 10^{-2}) = 19.9$ kg/m^{2}s.

At 325 K, $\mu = 0.0198$ mN s/m^2 or 1.98×10^{-5} N s/m^2

$$C_p = 1.0 \times 10^3 \text{ J/kg K}$$

Thus $\quad h_i \times 1.2 \times 10^{-2}/(2.9 \times 10^{-2}) = 0.023(1.2 \times 10^{-2} \times 19.9/1.98 \times 10^{-5})^{0.8}$
$$\times (1.0 \times 10^3 \times 1.98 \times 10^{-5}/0.029)^{0.4}$$
$$0.4138 h_i = 0.023(1.206 \times 10^4)^{0.8}(0.683)^{0.4}$$

and
$$h_i = 87.85 \text{ W/m}^2 \text{ K}$$

Outside:

In equation 7.68:
$$h_o d_o / k = 0.33 C_h (d_o G'/\mu)_{max}^{0.6} (C_p \mu / k)^{0.3}$$
$$d_o = 12.0 \text{ mm} \quad \text{or} \quad 1.2 \times 10^{-2} \text{ m}$$
$$G' = 10 \text{ kg/m}^2 \text{ s} \quad \text{for free flow}$$
$$G'_{max} = YG'/(Y - d_o)$$

where Y, the distance between tube centres $= 2d_o = 2.4 \times 10^{-2}$ m.

$\therefore \qquad G'_{max} = 2.4 \times 10^{-2} \times 10.0/(2.4 \times 10^{-2} - 1.2 \times 10^{-2}) = 20$ kg/m^2 s

At a mean flue gas temperature of $0.5(700 + 366) = 533$ K.

$$\mu = 0.0286 \text{ mN s/m}^2 \quad \text{or} \quad 2.86 \times 10^{-5} \text{ N s/m}^2$$
$$k = 0.045 \text{ W/m K}$$
$$C_p = 1.0 \times 10^3 \text{ J/kg K}$$

$\therefore \qquad Re_{max} = (1.2 \times 10^{-2} \times 20.0/2.86 \times 10^{-5}) = 8.39 \times 10^3$

From Table 7.3, when $Re_{max} = 8.39 \times 10^3$, $X = 2d_o$, and $Y = 2d_o$, $C_h = 0.95$.

Thus $\quad h_o \times 1.2 \times 10^{-2}/(4.5 \times 10^{-2}) = 0.33 \times 0.95(8.39 \times 10^3)^{0.6}$
$$\times (1.0 \times 10^3 \times 2.86 \times 10^{-5}/0.045)^{0.3}$$

or $\qquad 0.267 h_o = 0.314(8.39 \times 10^3)^{0.6}(0.836)^{0.3}$

and $\qquad h_o = 232$ W/m^2 K

Overall:

Ignoring wall and scale resistances:

$$1/U = 1/h_o + 1/h_i = (0.0114 + 0.0043) = 0.0157$$

and $\qquad U = 63.7$ W/m^2 K

Area required

In equation 7.1, $A = Q/U\theta_m = 74\cdot7 \times 10^3/(63\cdot7 \times 180) = 6\cdot52$ m².
Area/unit length of tube $= (\pi/4)(1\cdot2 \times 10^{-2}) = 9\cdot43 \times 10^{-3}$ m²/m and hence total length of tubing required $= 6\cdot52/(9\cdot43 \times 10^{-3}) = 6\cdot92 \times 10^2$ m.
The length of each tube is therefore $= 6\cdot92 \times 10^2/(20 \times 20)$
$$= \underline{\underline{1\cdot73 \text{ m}}}$$

Problem 7.5

A cooling coil, consisting of a single length of tubing through which water is circulated, is provided in a reaction vessel, the contents of which are kept uniformly at 360 K by means of a stirrer. The inlet and outlet temperatures of the cooling water are 280 K and 320 K respectively. What would the outlet water temperature become if the length of the cooling coil were increased 5 times? Assume the overall heat transfer coefficient to be constant over the length of the tube and independent of the water temperature.

Solution

Assuming that equation 7.1 may be used:

$$Q = UA\Delta T_m$$

where ΔT_m is the logarithmic mean temperature difference, then for the initial conditions,

$Q_1 = m_1 \times 4\cdot18(320 - 280) = U_1 A_1[(360 - 280) - (360 - 320)]/\ln(360 - 280)/(360 - 320)$
or $\qquad 167\cdot2 m_1 = U_1 A_1(80 - 40)/\ln 80/40 = 57\cdot7 U_1 A_1$
and $\qquad m_1/U_1 A_1 = 0\cdot345$

In the second case, $m_2 = m_1$, $U_2 = U_1$, and $A_2 = 5A_1$.

$\therefore\ Q_2 = m_1 \times 4\cdot18(T - 280) = 5U_1 A_1[(360 - 280) - (360 - T)]/\ln(360 - 280)/(360 - T)$
or $\qquad 4\cdot18(m_1/U_1 A_1)(T - 280)/5 = (80 - 360 + T)/[\ln 80/(360 - T)]$

Substituting for $m_1/U_1 A_1$,

$$0\cdot289(T - 280) = (T - 280)/[\ln 80/(360 - T)]$$
or $\qquad \ln 80/(360 - T) = 3\cdot467$
and $\qquad\qquad \underline{\underline{T = 357\cdot5 \text{ K}}}$

Problem 7.6

In an oil cooler, 60 g/s of hot oil enters a thin metal pipe of diameter 25 mm. An equal mass of cooling water flows through the annular space between the pipe and a larger concentric pipe, the oil and water moving in opposite directions. The oil enters at 420 K and is to be cooled to 320 K. If the water enters at 290 K, what length of pipe will be required? Take coefficients of 1·6 kW/m² K on the oil side and 3·6 kW/m² K on the water side and 2·0 kJ/kg K for the specific heat of the oil.

Solution

Heat load

Mass flow of oil = 60 g/s or $6·0 \times 10^{-2}$ kg/s.

and hence $$Q = 6·0 \times 10^{-2} \times 2·0(420 - 320) = 12 \text{ kW}$$

Thus the water outlet temperature is given by:

$$12 = 6·0 \times 10^{-2} \times 4·18(T - 290)$$

or $$T = 338 \text{ K}$$

Logarithmic mean temperature driving force

From equation 7.10:

$$\theta_1 = (420 - 338) = 82 \text{ K}, \quad \theta_2 = (320 - 290) = 30 \text{ K}$$

and $$\theta_m = (82 - 30)/\ln 82/30 = 51·7 \text{ K}$$

Overall coefficient

The pipe wall is thin and hence its thermal resistance may be neglected. Thus from equation 7.8:

$$1/U = 1/h_o + 1/h_i = (1/1·6 + 1/3·6)$$

and $$U = 1·108 \text{ kW/m}^2 \text{ K}$$

Area

In equation 7.9:

$$A = Q/U\theta_m = 12/(1·108 \times 51·7) = 0·210 \text{ m}^2$$

Tube diameter = 25×10^{-3} m (assuming a mean value)

and hence $\quad$ area/unit length = $(\pi \times 25 \times 10^{-3} \times 1·0) = 7·85 \times 10^{-2}$ m²/m

and hence $\quad$ tube length required = $0·210/(7·85 \times 10^{-2})$

$$= \underline{\underline{2·67 \text{ m}}}$$

Problem 7.7

The walls of a furnace are built up to 150 mm thickness of a refractory of thermal conductivity 1·5 W/m K. The surface temperatures of the inner and outer faces of the refractory are 1400 K and 540 K respectively.

If a layer of insulating material 25 mm thick of thermal conductivity 0·3 W/m K is added, what temperatures will its surfaces attain assuming the inner surface of the furnace to remain at 1400 K? The coefficient of heat transfer from the outer surface of the insulation to the surroundings, which are at 290 K, may be taken as 4·2, 5·0, 6·1, and 7·1 W/m K for surface temperatures of 370, 420, 470, and 520 K respectively. What will be the reduction in heat loss?

Solution

From equation 7.11:

heat flow through the refractory, $Q = kA(T_1 - T_2)/x$

Thus for unit area,

$$Q = 1·5 \times 1·0(1400 - T_2)/(150 \times 10^{-3}) = 14,000 - 10T_2 \text{ W/m}^2 \qquad \text{(i)}$$

where T_2 is the temperature at the refractory/insulation interface.

Similarly, the heat flow through the insulation is:

$$Q = 0·3 \times 1·0(T_2 - T_3)/(25 \times 10^{-3}) = 12T_2 - 12T_3 \text{ W/m}^2 \qquad \text{(ii)}$$

The flow of heat from the insulation surface at T_3 K to the surroundings at 290 K, is:

$$Q = hA(T_3 - 290) \quad \text{or} \quad hT_3 - 290h \text{ W/m}^2 \qquad \text{(iii)}$$

where h is the coefficient of heat transfer from the outer surface.

The solution is now made by trial and error as follows: a value of T_3 is selected and h obtained by interpolation of the given data. These are substituted in (iii) to give Q. T_2 is then obtained from (ii) and a second value of Q is then obtained from (i). The correct values of T_3 is then given when these two values of Q coincide. The working is as follows:

T_3 (K)	h (W/m² K)	$Q = h(T_3 - 290)$ (W/m²)	$T_2 = T_3 + Q/12$ (K)	$Q = 14,000 - 10T_2$ (W/m²)
300	3·2	32·0	302·7	10,973
350	3·9	234·0	369·5	10,305
400	4·7	517	443·1	9569
450	5·6	896	524·7	8753
500	6·45	1355	612·9	7871
550	7·8	2028	719·0	6810
600	9·1	2821	835·1	5649
650	10·4	3744	962·0	4380
700	11·5	4715	1092·9	3071
750	12·7	5842	1236·8	1632

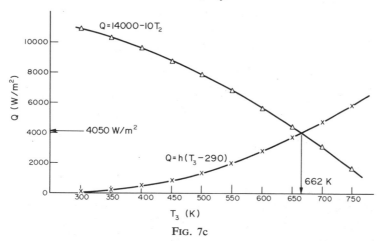

FIG. 7c

These data are plotted in Figure 7c from which a balance is obtained when $T_3 = 662$ K, at which $Q = 4050$ W/m².

In equation (i), $4050 = 14,000 - 10T_2$ or $T_2 = 995$ K

Thus the temperatures at the inner and outer surfaces of the insulation are

<u>995 K and 662 K respectively</u>

With no insulation, $Q = 1.5 \times 1.0(1400 - 540)/(150 \times 10^{-3})$
$= 8600$ W/m²

and hence the reduction heat loss is $(8600 - 4050) = \underline{4550 \text{ W/m}^2}$

or $(4540 \times 100)/8600 = \underline{52.9\%}$

Problem 7.8

A pipe of outer diameter 50 mm, maintained at 1100 K, is covered with 50 mm of insulation of thermal conductivity 0·17 W/m K.

Would it be feasible to use a magnesia insulation, which will not stand temperatures above 615 K and has a thermal conductivity of 0·09 W/m K, for an additional layer thick enough to reduce the outer surface temperature to 370 K in surroundings at 280 K? Take the surface coefficient of heat transfer by radiation and convection as 10 W/m² K.

Solution

For convection to the surroundings, $Q = hA_3(T_3 - T_4)$ W/m

where A_3 is area for heat transfer per unit length of pipe.

The radius of the pipe, $r_1 = 50/2 = 25$ mm or 0·025 m.
The radius of the insulation, $r_2 = (25 + 50) = 75$ mm or 0·075 m.

The radius of the magnesia, $r_3 = (75 + x) = 0 \cdot 075 + 0 \cdot 001x$ m where x mm is the thickness of the magnesia.

Hence the area at the surface of the magnesia, $A_3 = 2\pi(0 \cdot 075 + 0 \cdot 001x)$ m^2/m

and
$$Q = 10[2\pi(0 \cdot 075 + 0 \cdot 001x)](370 - 280)$$
$$= 424 \cdot 1 + 5 \cdot 66x \text{ W/m} \tag{i}$$

For conduction through the insulation

From equation 7.18, $\qquad Q = k(2\pi r_m l)(T_1 - T_2)/(r_2 - r_1)$

where $\qquad r_m = (r_2 - r_1)/\ln r_2/r_1$.

$\therefore \qquad Q = 0 \cdot 17[2\pi \times 1 \cdot 0(r_2 - r_1)](1100 - T_2)/[(r_2 - r_1)\ln(0 \cdot 075/0 \cdot 025)]$
$$= 0 \cdot 972(1100 - T_2) \text{ W/m} \tag{ii}$$

For conduction through the magnesia

In equation 7.18,

$$Q = 0 \cdot 09[2\pi \times 1 \cdot 0(r_3 - r_2)](T_2 - 370)/[(r_3 - r_2) \ln(0 \cdot 075 + 0 \cdot 001x)/0 \cdot 075]$$
$$= 0 \cdot 566(T_2 - 370)/\ln(1 + 0 \cdot 013x) \tag{iii}$$

For a selected value of x, Q is found from (i) and hence T_2 from (ii). These values are substituted in (iii) to give a second value of Q, with the following results:

x (mm)	$Q = 424 \cdot 1 + 5 \cdot 66x$ (W/m)	$T_2 = 1100 - 1 \cdot 028Q$ (K)	$Q = 0 \cdot 566(T_2 - 370)/\ln(1 + 0 \cdot 013x)$ (W/m)
5·0	452·4	635	2380
7·5	466·6	620	1523
10·0	480·7	606	1092
12·5	494·9	591	831·9
15·0	509·0	577	657·0
17·5	523·2	562	530·7
20·0	537·3	548	435·2

From a plot of the two values of Q a balance is attained when $x = 17 \cdot 5$ mm. With this thickness, $\underline{T_2 = 560 \text{ K}}$ which is below the maximum permitted and hence the use of the magnesia would be feasible.

Problem 7.9

In order to warm 0·5 kg/s of a heavy oil from 311 K to 327 K, it is passed through tubes of inside diameter 19 mm and length 1·5 m forming a bank, on the outside of which steam is condensing at 373 K. How many tubes will be needed?

In calculating Nu, Pr, and Re, the thermal conductivity of the oil may be taken as

0·14 W/m K and the specific heat as 2·1 kJ/kg K, irrespective of temperature. The viscosity is to be taken at the mean oil temperature. Viscosity of the oil at 319 and 373 K is 154 and 19·2 mN s/m² respectively.

Solution

Heat load
$$Q = 0.5 \times 2.1(327 - 311) = 16.8 \text{ kW}$$

Logarithmic mean driving force
$$\theta_1 = (373 - 311) = 62 \text{ K}, \quad \theta_2 = (373 - 327) = 46 \text{ K}$$

∴ From equation 7.10, $\quad\quad \theta_m = (62 - 46)/\ln(62/46) = 53.6 \text{ K}$

A preliminary estimate of the overall heat transfer coefficient may now be obtained from Table 7.13 as follows:

For condensing steam, $h_o = 10,000$ W/m² K; and for oil, $h_i = 250$ W/m² K (say). Thus $1/U = 1/h_o + 1/h_i = 0.0041$ and $U = 244$ W/m² K, and from equation 7.1 the preliminary area is:
$$A = 16.8 \times 10^3/(244 \times 53.6) = 1.29 \text{ m}^2$$

The area/unit length of tube is $\quad \pi \times 19.0 \times 10^{-3} \times 1.0 = 5.97 \times 10^{-2}$ m²/m

and $\quad\quad\quad\quad\quad$ total length of tubing $= 1.29/(5.97 \times 10^{-2}) = 21.5$ m

∴ $\quad\quad\quad\quad\quad$ number of tubes $= 21.5/1.5 = 14.3$, say 14 tubes

Film coefficients

The inside coefficient is controlling and hence this must be checked to ascertain if the preliminary estimate is valid.

The Reynolds number is given by $Re = d_i G'/\mu$
$$= 19.0 \times 10^{-3} G'/\mu$$

At a mean oil temperature of $0.5(327 + 311) = 319$ K, $\mu = 154 \times 10^{-3}$ N s/m².

Area for flow per tube $= (\pi/4)(19.0 \times 10^{-3})^2 = 2.835 \times 10^{-4}$ m².

∴ $\quad\quad\quad\quad$ total area for flow $= 14 \times 2.835 \times 10^{-4} = 3.969 \times 10^{-3}$ m²

and hence $\quad\quad\quad G' = 0.5/(3.969 \times 10^{-3}) = 1.260 \times 10^2$ kg/m²s

Thus $\quad\quad\quad\quad Re - 19.0 \times 10^{-3} \times 1.260 \times 10^2/(154 \times 10^{-3}) = 15.5$

That is, the flow is streamline and equation 7.63 may be used:
$$(h_i d_i/k)(\mu_s/\mu)^{0.14} = 2.01(GC_p/kl)^{0.33}$$

At a mean wall temperature of $0.5(373 + 319) = 346$ K, $\mu_s = 87.0 \times 10^{-3}$ N s/m². The mass flow, $G = 0.5$ kg/s.

∴ $\quad (h_i \times 19.0 \times 10^{-3}/0.14)(87.0 \times 10^{-3}/154 \times 10^{-3})^{0.14}$
$$= 2.01(0.5 \times 2.1 \times 10^3/0.14 \times 1.5)^{0.33}$$

or $\quad\quad\quad\quad\quad 0.136 h_i \times 0.923 = 2.01 \times 16.6$

∴ $\quad\quad\quad\quad\quad\quad\quad h_i = 266$ W/m² K

This is sufficiently close to the assumed value and hence 14 tubes would be specified.

Problem 7.10

A metal pipe of 12 mm outer diameter is maintained at 420 K. Calculate the rate of heat loss per metre run in surroundings uniformly at 290 K, (a) when the pipe is covered with 12 mm thickness of a material of thermal conductivity 0·35 W/mK and surface emissivity 0·95, and (b) when the thickness of the covering material is reduced to 6 mm, but the outer surface is so treated so as to reduce its emissivity to 0·10.

The coefficients of radiation from a perfectly black surface in surroundings at 290 K are 6·25, 8·18, and 10·68 W/m² K at 310 K, 370 K, and 420 K respectively. The coefficients of convection may be taken as $1·22(\theta/d)^{0·25}$ W/m² K, where θ(K) is the temperature difference between the surface and the surrounding air and d(m) is the outer diameter.

Solution

Case (a)

Assume that the heat loss is q W/m and the surface temperature is T K. Thus for conduction through the insulation, from 7.11:

$$q = kA_m(420 - t)/x$$

The mean diameter = 18 mm or 0·018 m, and hence

$$A_m = (\pi \times 0·018) = 0·0566 \text{ m}^2/\text{m} \qquad x = 0·012 \text{ m}$$

$$\therefore \qquad q = 0·35 \times 0·0566(420 - T)/0·012 = 693·3 - 1·67T \text{ W/m} \qquad \text{(i)}$$

For convection and radiation from the surface, from equation 7.86:

$$q = (h_r + h_c)A_2(T - 290) \text{ W/m}$$

where h_r is the film coefficient equivalent to the radiation and h_c the coefficient due to convection given by:

$$h_c = 1·22[(T - 290)/d]^{0·25} \quad \text{where } d = 36 \text{ mm or } 0·036 \text{ m}$$

$$\therefore \qquad h_c = 2·80(T - 290)^{0·25} \text{ W/m}^2 \text{ K}$$

If h_b is the coefficient equivalent to radiation from a black body,

$$h_r = 0·95h_b \text{ W/m}^2 \text{ K}$$

The outer diameter = 0·036 m and hence

$$A_2 = (\pi \times 0·036 \times 1·0) = 0·1131 \text{ m}^2/\text{m}$$

$$\therefore \qquad q = [0·95h_b + 2·80(T - 290)^{0·25}]0·1131(T - 290)$$

$$= 0·1074h_b(T - 290) + 0·317(T - 290)^{1·25} \text{ W/m} \qquad \text{(ii)}$$

Values of T are now assumed and together with values of h_b from the given data substituted into (i) and (ii) until equal values of q are obtained as follows:

T (K)	$q = 693 \cdot 3 - 1 \cdot 67T$ (W/m)	h_b (W/m² K)	$0 \cdot 1074 h_b (T - 290)$ (W/m)	$0 \cdot 317 (T - 290)^{1 \cdot 25}$ (W/m)	q (W/m)
300	193·3	6·0	6·5	5.7	12·2
320	160·0	6·5	20·9	22·2	43·1
340	126·7	7·1	38·1	42·1	80·2
360	93·3	7·8	58·7	64·2	122·9
380	60·0	8·55	82·7	87·9	170·6
400	26·7	9·55	112·8	113·0	225·8

A balance is obtained when $T = 352\ \text{K}$ and $q = 106\ \text{W/m}$.

Case (b)

For conduction through the insulation, $x = 0 \cdot 006\ \text{m}$ and the mean diameter $= 15\ \text{mm}$ or $0 \cdot 015\ \text{m}$.

$$\therefore \qquad A_m = \pi \times 0 \cdot 015 \times 1 \cdot 0 = 0 \cdot 0471\ \text{m}^2/\text{m}$$

$$\therefore \qquad q = 0 \cdot 35 \times 0 \cdot 0471 (420 - T)/0 \cdot 006$$

$$= (1154 - 2 \cdot 75T)\ \text{W/m} \qquad (\text{i})$$

The outer diameter is now $0 \cdot 024\ \text{m}$ and therefore

$$A_2 = (\pi \times 0 \cdot 024 \times 1 \cdot 0) = 0 \cdot 0754\ \text{m}^2/\text{m}$$

The coefficient due to convection is:

$$h_c = 1 \cdot 22 [(T - 290)/0 \cdot 024]^{0 \cdot 25} = 3 \cdot 10 (T - 290)^{0 \cdot 25}\ \text{W/m}^2\ \text{K}$$

The emissivity is $0 \cdot 10$ and hence $\qquad h_r = 0 \cdot 10 h_b\ \text{W/m}^2\ \text{K}$

$$\therefore \qquad q = [0 \cdot 10 h_b + 3 \cdot 10 (T - 290)^{0 \cdot 25}] 0 \cdot 0754 (T - 290)$$

$$= 0 \cdot 00754 h_b (T - 290) + 0 \cdot 234 (T - 290)^{1 \cdot 25}\ \text{W/m} \qquad (\text{ii})$$

Making the calculation as before:

T (K)	$q = 1154 - 2 \cdot 75T$ (W/m)	h_b (W/m² K)	$0 \cdot 00754 h_b (T - 290)$ (W/m)	$0 \cdot 234 (T - 290)^{1 \cdot 25}$ (W/m)	q (W/m)
300	329·0	6·0	0·5	4·2	4·7
320	274·0	6·5	1·5	16·4	17·9
340	219·0	7·1	2·7	31·1	33·8
360	164·0	7·8	4·2	47·4	51·6
380	109·0	8·55	5·8	64·9	70·7
400	54·0	9·55	7·9	83·4	91·3

A balance is obtained when $T = 390\ \text{K}$ and $q = 81\ \text{W/m}$.

Problem 7.11

A condenser consists of 30 rows of parallel pipes of outer diameter 230 mm and thickness 1·3 mm with 40 pipes, each 2 m long per row. Water, inlet temperature 283 K, flows through the pipes at 1 m/s and steam at 372 K condenses on the outside of the pipes. There is a layer of scale 0·25 mm thick of thermal conductivity 2·1 W/m K on the inside of the pipes.

Taking the coefficients of heat transfer on the water side as 4.0 and on the steam side as 8·5 kW/m² K, calculate the water outlet temperature and the total weight of steam condensed per second. The latent heat of steam at 372 K is 2250 kJ/kg. 1 m³ water weighs 1000 kg.

Solution

Overall coefficient

From equation 7.130:

$$\frac{1}{U} = \frac{1}{h_i} + \frac{1}{h_0} + \frac{x_w}{k_w} + \frac{x_r}{k_r}$$

where x_r and k_r are the thickness and thermal conductivity of the scale. Considering these in turn, $h_i = 4000$ W/m² K.

The inside diameter, $d_i = 230 - (2 \times 1·3) = 227·4$ mm or 0·2274 m.

Therefore basing the coefficient on the outside diameter,

$$h_{io} = (4000 \times 0·2274/0·230) = 3955 \text{ W/m}^2 \text{ K}$$

For conduction through the wall, $x_w = 1·3$ mm, and from Table 7.1, k_w (for steel) = 45 W/m K.

$$\therefore \qquad\qquad k_w/x_w = (45/0·0013) = 34615 \text{ W/m}^2 \text{ K}$$

The mean wall diameter = $(0·230 + 0·2274)/2 = 0·2287$ m and hence the coefficient equivalent to the wall resistance based on the tube o.d. is:

$$34615 \times 0·2287/0·230 = 34419 \text{ W/m}^2 \text{ K}$$

For conduction through the scale, $x_r = 0·25 \times 10^{-3} m$, $k_r = 2·1$ W/m K

$$\therefore \qquad\qquad k_r/x_r = 2·1/0·25 \times 10^{-3} = 8400 \text{ W/m}^2 \text{ K}$$

The mean scale diameter = $(227·4 - 0·25) = 227·15$ mm or 0·2272 m and hence the coefficient equivalent to the scale resistance based on the tube o.d. is:

$$8400 \times 0·2272/0·230 = 8298 \text{ W/m}^2 \text{ K}$$

$$\therefore \qquad 1/U = 1/3955 + 1/8500 + 1/34419 + 1/8298 = 5·201 \times 10^{-4}$$

and $\qquad\qquad U = 1923$ W/m² K

Temperature driving force

If water leaves the unit at T K:

$$\theta_1 = (372 - 283) = 89 \text{ K}$$
$$\theta_2 = (372 - T)$$

and in equation 7.10:

$$\theta_m = [89 - (372 - T)]/\ln[89/(372 - T)]$$
$$= (T - 283)/\ln[89/(372 - T)]$$

Area

For 230 mm o.d. tubes, outside area per unit length $= (\pi \times 0 \cdot 230 \times 1 \cdot 0) = 0 \cdot 723 \text{ m}^2/\text{m}$.

Total length of tubes $= (30 \times 40 \times 2) = 2400 \text{ m}$
and hence heat transfer area $A = (2400 \times 0 \cdot 723) = 1735 \cdot 2 \text{ m}^2$

Heat load

The cross area for flow/tube $= (\pi/4)(0 \cdot 230)^2 = 0 \cdot 0416 \text{ m}^2/\text{tube}$.

Assuming a single-pass arrangement, there are 1200 tubes per pass, and hence area for flow $= (1200 \times 0 \cdot 0416) = 49 \cdot 86 \text{ m}^2$.

For a velocity of $1 \cdot 0$ m/s, the volumetric flow $= (0.1 \times 49 \cdot 86) \text{ m}^3/\text{s}$ and the mass flow $= (1000 \times 4 \cdot 986) = 4986 \text{ kg/s}$.

Thus the heat load, $Q = 4986 \times 4 \cdot 18(T - 283) = 20,840(T - 283) \text{ kW}$ or $2 \cdot 084 \times 10^7(T - 283) \text{ W}$.

Substituting for Q, U, A, and θ_m in equation 7.9:

$$2 \cdot 084 \times 10^7(T - 283) = 1923 \times 1735 \cdot 2(T - 283)/\ln[89/(372 - T)]$$

or
$$\ln[89/(372 - T)] = 0 \cdot 1601$$
and
$$\underline{T = 296 \text{ K}}$$

The total heat load is, therefore, $Q = 20,840(296 - 283) = 2709 \times 10^5 \text{ kW}$.
The weight of steam condensed $= 2 \cdot 709 \times 10^5/2250 = \underline{\underline{120 \cdot 4 \text{ kg/s}}}$.

Problem 7.12

In an oil cooler, water flows at the rate of 100 g/s per tube through metal tubes of outer diameter 19 mm and thickness 1·3 mm, along the outside of which oil flows in the opposite direction at the rate of 75 g/s per tube.

If the tubes are 2 m long and the inlet temperatures of the oil and water are respectively 370 K and 280 K, what will be the outlet oil temperature? The coefficient of heat transfer on the oil side is 1·7 and on the water side 2·5 kW/m² K and the specific heat of the oil is 1·9 kJ/kg K.

Solution

In the absence of data as to the geometry of the unit, the solution will be worked on the basis of *one tube*—a valid approach as the number of tubes effectively appears on both sides of equation 7.9.

Let T_w and T_o be the outlet temperatures of the water and the oil respectively.

Heat load

$$Q = 0 \cdot 100 \times 4 \cdot 18(T_w - 280) = 0 \cdot 418(T_w - 280) \text{ kW} \quad \text{for water}$$

and

$$Q = 0 \cdot 075 \times 1 \cdot 9(370 - T_o) = 0 \cdot 143(370 - T_o) \text{ kW} \quad \text{for the oil.}$$

From these two equations: $\qquad T_w = 406 \cdot 5 - 0 \cdot 342 T_o \text{ K}$

Area

For 19·0 mm o.d. tubes, $\qquad$ area $= (\pi \times 0 \cdot 019 \times 1 \cdot 0) = 0 \cdot 0597 \text{ m}^2/\text{m}$

and for one tube, $\qquad$ area $= (2 \cdot 0 \times 0 \cdot 0597) = 0 \cdot 1194 \text{ m}^2$

Temperature driving force

$$\theta_1 = (370 - T_w), \quad \theta_2 = (T_o - 280)$$

and

$$\theta_m = [(370 - T_w) - (T_o - 280)]/[\ln(370 - T_w)/(T_o - 280)]$$

$$= (650 - T_w - T_o)/\ln(370 - T_w)/(T_o - 280)$$

Substituting for t_w,

$$\theta_m = (243 \cdot 5 - 0 \cdot 658 T_o)/\ln(0 \cdot 342 T_o - 36 \cdot 5)/(T_o - 280) \text{ K}$$

Overall coefficient

$$h_i = 2 \cdot 5 \text{ kW/m}^2 \text{ K}$$

$$d_i = 19 \cdot 0 - (2 \times 1 \cdot 3) = 16 \cdot 4 \text{ mm}$$

Therefore the inside coefficient, based on the outside diameter,

$$h_{io} = 2 \cdot 5 \times 16 \cdot 4/19 \cdot 0 = 2 \cdot 16 \text{ kW/m}^2 \text{ K}$$

Neglecting the scale and wall resistances,

$$1/U = 1/2 \cdot 16 + 1/1 \cdot 7 = 1 \cdot 052 \text{ m}^2 \text{ K/kW}$$

and $\qquad U = 0 \cdot 951 \text{ kW/m}^2 \text{ K}$

Therefore, substituting in equation 7.9:

$$0 \cdot 143(370 - T_o) = 0 \cdot 951 \times 0 \cdot 1194(243 \cdot 5 - 0 \cdot 658 T_o)/\ln(0 \cdot 342 T_o - 36 \cdot 5)/(T_o - 280)$$

$\therefore \qquad \ln(0 \cdot 342 T_o - 36 \cdot 5)/(T_o - 280) = 0 \cdot 523$

and $\qquad \underline{\underline{T_o = 324 \text{ K}}}$

Problem 7.13

Waste gases flowing across the outside of a bank of pipes are being used to warm air which flows through the pipes. The bank consists of 12 rows of pipes with 20 pipes, each 0·7 m long, per row. They are arranged in-line, with centre-to-centre spacing equal in both directions to one-and-a-half times the pipe diameter. Both inner and outer diameter may be taken as 12 mm. Air with a mass velocity of 8 kg/m² s enters the pipes at 290 K. The initial gas temperature is 480 K and the total weight of the gases crossing the pipes per second is the same as the total weight of the air flowing through them.

Neglecting gas radiation, estimate the outlet temperature of the air. The physical constants for the waste gases may be assumed the same as for air, given below:

Temperature (K)	Thermal conductivity (W/m K)	Viscosity (mN s/m²)
250	0·022	0·0165
310	0·027	0·0189
370	0·030	0·0214
420	0·033	0·0239
480	0·037	0·0260

Specific heat = 1·00 kJ/kg K.

Solution

Heat load

The cross area for flow per pipe = $(\pi/4)(0·012)^2 = 0·000113$ m² and therefore for $(12 \times 20) = 240$ pipes, flow area = $(240 \times 0·000113) = 0·0271$ m².

Thus,   flow of air = $(8·0 \times 0·271) = 0·217$ kg/s

which is also equal to the flow of waste gas.

If the outlet temperatures of the air and waste gas are T_a and T_w K respectively, then

$$Q = 0·217 \times 1·0(T_a - 290) \text{ kW} \quad \text{or} \quad 217(T_a - 290) \text{ W}$$

and

$$Q = 0·217 \times 1·0(480 - T_w) \text{ kW}$$

from which

$$T_w = (770 - T_a) \text{ K}$$

Area

Surface area/unit length of pipe = $(\pi \times 0·012 \times 1·0) = 0·0377$ m²/m.

Total length of pipe = $(240 \times 0·7) = 168$ m

and hence the heat transfer area, $A = (168 \times 0·0377) = 6·34$ m².

Temperature driving force

$$\theta_1 = (480 - T_a)$$
$$\theta_2 = (T_w - 1290)$$

or, substituting for t_w, $\theta_2 = (480 - T_a) = \theta_1$

∴ $\theta = (480 - T_a)$

Overall coefficient

The solution is now one of trial and error in that mean temperatures of both streams must be assumed in order to evaluate the physical properties.

Inside the tubes:
a mean temperature of 320 K, will be assumed at which,

$k = 0.028$ W/m K, $\mu = 0.0193 \times 10^{-3}$ N s/m^2,   and $C_p = 1.0 \times 10^3$ J/kg K

Therefore in equation 7.50,

$$h_i d_i / k = 0.023 (d_i G/\mu)^{0.8} (C_p \mu/k)^{0.4}$$
$$(h_i \times 0.012/0.028) = 0.023(0.012 \times 8.0/0.0193 \times 10^{-3})^{0.8}(1 \times 10^3$$
$$\times 0.0193 \times 10^{-3}/0.028)^{0.4}$$
$$h_i = 0.0537(4.974 \times 10^3)^{0.8}(0.689)^{0.4}$$
$$= 41.94 \text{ W/m}^2 \text{ K}$$

Outside the tubes:

The cross area of the tube bundle $= 0.7 \times 20(1.5 \times 0.012) = 0.252$ m^2 and hence the free flow mass velocity, $G' = 0.217/0.252 = 0.861$ kg/m^2 s.

From figure 7.20, $Y = (1.5 \times 0.012) = 0.018$ m

and ∴ $G'_{max} = (0.861 \times 0.018)/(0.018 - 0.012) = 2.583$ kg/m^2 s

At an assumed mean temperature of 450 K, $\mu = 0.0250 \times 10^{-3}$ N s/m^2 and $k = 0.035$ W/m K.

∴ $Re_{max} = 0.012 \times 2.583/(0.0250 \times 10^{-3}) = 1.24 \times 10^3$

Therefore, from Table 7.3,   for $X = 1.5d_o$ and $Y = 1.5d_o$, $C_h = 0.95$

In equation 7.68;

$$h_o 0.012/0.035 = 0.33 \times 0.95(1.24 \times 10^3)^{0.6}(1.0 \times 10^3 \times 0.0250 \times 10^{-3}/0.035)^{0.3}$$
∴ $$h_o = 0.914 \times 71.8 \times 0.714^{0.3} = 59.3 \text{ W/m}^2 \text{ K}$$

Hence, ignoring wall and scale resistances,

$$1/U = 1/41.94 + 1/59.3 = 4.07 \times 10^{-2}$$

and $U = 24.57$ W/m^2 K

Thus in equation 7.9:

$$217(T_a - 290) = 24.57 \times 6.34(480 - T_a)$$

from which $\underline{\underline{T_a = 369.4 \text{ K}}}$

With this value the mean air and waste gas temperatures are 330 K and 440 K respectively. These are within 10 K of the assumed values in each case and the error would have a negligible effect on the film properties rendering recalculation unnecessary.

Problem 7.14

Oil is to be warmed from 300 K to 344 K by passing it at 1 m/s through the pipes of a shell-and-tube heat exchanger. Steam at 377 K condenses on the outside of the pipes, which have outer and inner diameters of 48 and 41 mm respectively; but, owing to fouling, the inside diameter has been reduced to 38 mm, and the resistance to heat transfer of the pipe wall and dirt together, based on this diameter, is $0 \cdot 0009 \, m^2 \, K/W$.

It is known from previous measurements under similar conditions that the oil side coefficients of heat transfer for a velocity of 1 m/s, based on a diameter of 38 mm, vary with the temperature of the oil according to the table below:

Oil temperature (K)	300	311	322	333	344
Oil side coefficient of heat transfer (W/m² K)	74	80	97	136	244

The specific heat and density of the oil may be assumed constant at 1·9 kJ/kg K and 900 kg/m³ respectively, and any resistance to heat transfer on the steam side neglected.

Find the length of tube bundle required.

Solution

This problem, in the absence of further data, may be worked on the basis of one tube.

Heat load

Cross-sectional area at the inside diameter of the scale $= (\pi/4)(0 \cdot 038)^2 = 0 \cdot 00113 \, m^2$.

∴ $\qquad$ volumetric flow $= (0 \cdot 00113 \times 1 \cdot 0) = 0 \cdot 00113 \, m^3/s$

and $\qquad$ mass flow $= (0 \cdot 00113 \times 900) = 1 \cdot 021 \, kg/s$

∴ $\qquad$ heat load, $Q = 1 \cdot 021 \times 1 \cdot 9(344 - 300) = 85 \cdot 33 \, kW$

Temperature driving force

$$\theta_1 = (377 - 300) = 77 \, K, \quad \theta_2 = (377 - 344) = 33 \, K$$

and $\qquad$ $\theta_m = (77 - 33)/\ln(77/33) = 52 \, K$

Overall coefficient

Inside:

The mean oil temperature $= 0.5(344 + 300) = 322$ K at which h_i (based on $d_i = 0.038$ m) $= 97$ W/m^2 K.

Basing this value on the outside diameter of the pipe,

$$h_{io} = 97 \times 0.038/0.048 = 76.8 \text{ W/m}^2 \text{ K}$$

Outside:

From table 7.13, h_o for condensing steam will be taken as 10,000 W/m^2 K.
Wall and scale:

The scale resistance based on $d = 0.038$ m is 0.0009 m^2 K/W or $k/x = 1/0.0009 = 1111.1$ W/m^2 K.

Basing this on the tube o.d., $k/x = 1111.1 \times 0.038/0.048 = 879.6$ W/m^2 K.

$\therefore$
$$1/U = 1/h_{io} + 1/h_o + x/k$$
$$= 0.0130 + 0.0001 + 0.00114$$

or
$$U = 70.2 \text{ W/m}^2 \text{ K}$$

Area

$$A = Q/U\theta_m = 85.33 \times 10^3/(70.2 \times 52) = 23.4 \text{ m}^2$$

Area per unit length of pipe $\qquad = \pi \times 0.048 \times 1.0 = 0.151 \text{ m}^2/\text{m}$

and $\qquad$ length of tube bundle $= 23.4/0.151 = \underline{\underline{154.9 \text{ m}}}$

This very large tube length is a reflection of the very low inside film coefficient and such a result would necessitate the use of several passes or indeed a multistage unit. A better approach would be to increase the tube side velocity by decreasing the number of tubes per pass, though any pressure drop limitations would have to be taken into account. Further, the use of a smaller tube diameter should be considered.

Problem 7.15

It is proposed to construct a heat exchanger to condense 7.5 kg/s of n-hexane at a pressure of 150 kN/m^2, involving a heat load of 4.5 MW. The hexane is to reach the condenser from the top of a fractionating column at its condensing temperature of 356 K.

From experience it is anticipated that the overall heat transfer coefficient will be 450 W/m^2 K. The available cooling water is at 289 K.

Outline the proposals that you would make for the type and size of the exchanger, and explain the details of the mechanical construction that you consider require special attention.

Solution

A shell-and-tube unit is suitable with hexane on the shell side. For a heat load of $4 \cdot 5$ MW $= 4 \cdot 5 \times 10^3$ kW, the outlet water temperature is given by:

$$4 \cdot 5 \times 10^3 = m \times 4 \cdot 18(t - 289)$$

In order to avoid severe scaling in such a case, the maximum allowable water temperature is 320 K and hence 310 K will be chosen as a suitable value for t.

Thus $\qquad 4 \cdot 5 = 10^3 = 4 \cdot 18m(310 - 289) \quad$ and $\quad m = 51 \cdot 3$ kg/s

The next stage is to estimate the heat transfer area required using equation 7.9:

$$Q = UA\theta_m$$

where the heat load $Q = 4 \cdot 5 \times 10^6$ W
$$U = 450 \text{ W/m}^2 \text{ K}$$

$$\theta_1 = (356 - 289) = 67 \text{ K}, \quad \theta_2 = (356 - 310) = 46 \text{ K}$$

and from equation 7.10: $\qquad \qquad \theta_m = 67 - 46/\ln(67/46) = 55 \cdot 8$ K

No correction factor is necessary as the shell side fluid temperature is constant.

$\therefore \qquad \qquad A = 4 \cdot 5 \times 10^6/(450 \times 55 \cdot 8) = 179 \cdot 2 \text{ m}^2$

A reasonable tube size must now be selected, say $25 \cdot 4$ mm, 14 BWG.
The outside surface area is therefore $(\pi \times 0 \cdot 0254 \times 1 \cdot 0) = 0 \cdot 0798 \text{ m}^2/\text{m}$ and hence the total length of tubing required $= (179 \cdot 2/0 \cdot 0798) = 2246$ m.
A standard tube length is now selected, say $4 \cdot 87$ m and hence the total number of tubes required $= (2246/4 \cdot 87) = 460$.
It now remains to decide the number of tubes/pass, which is obtained from consideration of the water velocity. For shell and tube units, $u = 1 \cdot 0 - 1 \cdot 5$ m/s and a value of $1 \cdot 25$ m/s will be selected.
The water flow, $51 \cdot 3$ kg/s $= (51 \cdot 3/1000) = 0 \cdot 0513 \text{ m}^3/\text{s}$.
The tube i.d. is $21 \cdot 2$ mm and hence the cross-sectional area for flow/tube $= (\pi/4)(0 \cdot 0212)^2 = 0 \cdot 000353 \text{ m}^2$.
Area required to give a velocity of $1 \cdot 25$ m/s $= (0 \cdot 0513/1 \cdot 25) = 0 \cdot 0410 \text{ m}^2$
and hence number of tubes/pass $= (0 \cdot 0410/0 \cdot 000353) = 116$
and number of passes $= 460/116 \approx 4$.
With the shell side fluid being clean (hexane), triangular pitch might be suitable and 460×25 mm o.d. tubes on 32 mm triangular pitch with 4 tube side passes can be accommodated in a $0 \cdot 838$ m i.d. shell and still allow room for impingement plates.
The proposed unit will therefore consist of:
460, $25 \cdot 4$ mm o.d. tubes $\times$ 14 BWG, $4 \cdot 87$ m long arranged with 4 tube side passes on 32 mm triangular pitch in a $0 \cdot 838$ m i.d. shell.
The general mechanical details of the unit are described in Section 7.9 and points of detail are:

(i) impingement baffles should be fitted under each inlet nozzle;

(ii) segmental baffles are not usually fitted to a condenser of this type;
(iii) the unit should be installed on saddles at say 5° to the horizontal to facilitate
drainage of the condensate.

Problem 7.16

A heat exchanger is to be mounted at the top of a fractionating column about
15 m high to condense 4 kg/s of n-pentane at 205 kN/m², corresponding to a
condensing temperature of 333 K. Give an outline of the calculations you would
make to obtain an approximate idea of the size and construction of the exchanger
required.

For purposes of standardisation, the company will use 19 mm outer diameter
tubes of 1·65 mm wall thickness and these may be 2·5, 3·6, or 5 m in length. The
film coefficient for condensing pentane on the outside of a horizontal tube bundle
may be taken as 1·1 kW/m² K. The condensation is effected by pumping water
through the tubes, the initial water temperature being 288 K.

The latent heat of condensation of pentane is 335 kJ/kg.

For these 19 mm tubes, a water velocity of 1 m/s corresponds to a flowrate of
200 g/s of water.

Solution

The calculations follow the sequence of previous problems in that heat load,
temperature driving force, and overall coefficient are obtained and hence the area
evaluated. It then remains to consider the geometry of the unit bearing in mind the
need to maintain a reasonable cooling water velocity.

As in the previous example, the n-pentane will be passed through the shell and
cooling water through the tubes.

Heat load

$Q = 4·0 \times 335 = 1340$ kW assuming there is no sub-cooling of the condensate.

As in Problem 7.15, the outlet temperature of the cooling water will be taken as
310 K, and for a flow of m kg/s,

$$1340 = m \times 4·18(310 - 288) \quad \text{or} \quad m = 14·57 \text{ kg/s}$$

Temperature driving force

$$\theta_1 = (333 - 288) = 45 \text{ K}, \quad \theta_2 = (333 - 310) = 23 \text{ K}$$

and

$$\theta_m = (45 - 23)/\ln(45/23) = 32·8 \text{ K}$$

Overall coefficient

Inside:

For forced convection to water in tubes, equation 7.59 may be used:

$$h_i = 1063(1 + 0.00293\,T)u^{0.8}/d_i^{0.2}$$

where T, the mean water temperature $= 0.5(310 + 288) = 299$ K; u, the water velocity will be taken as 1 m/s—a realistic optimum value, bearing in mind the need to limit the pressure drop, and $d_i = (19.0 - 2 \times 1.65) = 15.7$ mm or 0.0157 m.

$$\therefore \qquad h_i = 1063(1 + 0.00293 \times 299)1.0^{0.8}/0.0157^{0.2}$$
$$= 1063 \times 1.876 \times 1.0/0.436 = 4574 \text{ W/m}^2 \text{ K}$$

or, based on the outer diameter
$$h_{io} = 4.574 \times 0.0157/0.019$$
$$= 3.78 \text{ kW/m}^2 \text{ K}$$

Wall:

For steel, $k = 45$ W/m K and $x = 0.00163$ m

and hence $\qquad x/k = 0.00163/45 = 0.0000362 \text{ m}^2\text{K/W} \quad$ or $\quad 0.0362 \text{ m}^2\text{K/kW}$

Outside:

$$h_o = 1.1 \text{ kW/m}^2 \text{ K}$$

$\therefore \qquad 1/U = 1/h_o + x/k + 1/h_{io} \quad$ ignoring the scale resistance
$$= 0.9091 + 0.0362 + 0.2646$$

and $\qquad U = 0.827 \text{ kW/m}^2 \text{ K}$

Area

$$Q = UA\theta_m$$

and hence $\qquad A = 1340/(0.827 \times 32.8) = 49.4 \text{ m}^2$

Outer area of 0.019 m diameter tube $= (\pi \times 0.019 \times 1.0) = 0.0597 \text{ m}^2/\text{m}$ and hence total length of tubing required $= (49.4/0.0597) = 827.6$ m.

Thus with 2.5, 3.6, and 5.0 m tubes, the number of tubes will be 331, 230 or 166.

The total cooling water flow $= 14.57$ kg/s

and for $u = 1$ m/s, the flow through 1 tube is 0.20 kg/s

$\therefore \qquad$ the number of tubes/pass $= (14.57/0.20) = 73$

Clearly 3 passes are out of the question, and hence 2 or 4 are suitable, i.e. 146 or 292 tubes, 5.0 or 2.5 m long.

The former is closer to a standard shell size and 166×19 mm tubes on 25.4 mm square pitch with two tube side passes can be fitted within a 438 mm i.d. shell. In this event, the water velocity would be slightly less than 1 m/s (in fact $1 \times 146/166 = 0.88$ m/s), though this would not affect the overall coefficient to any significant extent.

The proposed unit is therefore 166 19 mm o.d. tubes on 25.4 mm square pitch 5.0 m long with a 438 mm i.d. shell.

In making such calculations it is good practice to add an overload factor to the heat load, say 10%, to allow for errors in predicting film coefficients, although this is often taken into account in allowing for extra tubes within the shell. In this particular example, the fact that the unit is to be installed 15 m above ground level is of significance in limiting the pressure drop and it may be that in a real situation space limitations would immediately specify the tube length.

Problem 7.17

An organic liquid is boiling at 340 K on the inside of a metal surface of thermal conductivity 42 W/m K and thickness 3 mm. The outside of the surface is heated by condensing steam. Assuming that the heat transfer coefficient from steam to the outer metal surface is constant at 11 kW/m² K, irrespective of the steam temperature, find what value of the steam temperature would give a maximum rate of evaporation.

The coefficients of heat transfer from the inner metal surface to the boiling liquid depend upon the temperature difference as shown below:

Temperature difference metal surface to boiling liquid (K)	Heat transfer coefficient metal surface to boiling liquid (kW/m² K)
22·2	4·43
27·8	5·91
33·3	7·38
36·1	7·30
38·9	6·81
41·7	6·36
44·4	5·73
50·0	4·54

Solution

For a steam temperature T_s K, the heat conducted through the condensing steam film, $Q = h_c A(t_s - t_1)$

or
$$q = 11 \times 1 \cdot 0(T_s - T_1) = 11 \cdot 0(T_s - T_1) \text{ kW/m}^2 \qquad \text{(i)}$$

where t_1 is the temperature of the outer surface of the metal.
For conduction through the metal,

$$q = kA(T_1 - T_2)/x$$
$$= 42 \times 10^{-3} \times 1 \cdot 0(T_1 - T_2)/0 \cdot 003 = 14 \cdot 0(T_1 - T_2) \text{ kW/m}^2 \qquad \text{(ii)}$$

where T_2 is the temperature of the inner surface of the metal.
For conduction through the boiling film:

$$q = h_b A(T_2 - 340) = h_b(T_2 - 340) \text{ kW/m}^2 \qquad \text{(iii)}$$

where h_b kW/m² K is the film coefficient to the boiling liquid.

Thus for a value of T_2 the temperature difference $(T_2 - 340)$ is obtained and h_b from the table of data. q is then obtained from (iii), T_1 from (ii), and hence T_s from (i) as follows:

T_2 (K)	$(T_2 - 340)$ (K)	h_b (kW/m² K)	q (kW/m²)	T_1 (K)	T_s (K)
362·2	22·2	4·43	98·4	369·2	378·1
367·8	27·8	5·91	164·3	379·5	394·4
373·3	33·3	7·38	245·8	390·9	413·3
376·1	36·1	7·30	263·5	394·9	418·9
378·9	38·9	6·81	264·9	397·8	421·9
381·7	41·7	6·36	265·2	400·7	424·8
384·4	44·4	5·73	254·4	402·6	425·7
390·0	50·0	4·54	227·0	406·2	426·8

It is fairly obvious that the rate of evaporation will be highest when the heat flux is a maximum. On inspection this occurs when $\underline{\underline{T_s = 425 \text{ K}}}$.

Problem 7.18

It is desired to warm an oil of specific heat 2·0 kJ/kg K from 300 K to 325 K by passing it through a tubular heat exchanger with metal tubes of inner diameter 10 mm. Along the outside of the tubes flows water, inlet temperature 372 K, and outlet temperature 361 K.

The overall heat transfer coefficient from water to oil, reckoned on the inside area of the tubes, may be assumed constant at 230 W/m² K, and 75 g/s of oil is to be passed through each tube.

The oil is to make two passes through the heater. The water makes one pass along the outside of the tubes. Calculate the length of the tubes required.

Solution

Heat load

If the total number of tubes is n, there are $n/2$ tubes in one pass on the oil side. In other words the oil passes through 2 tubes in traversing the exchanger.

The mass flow of oil is therefore $= 75 \times n/2 = 37\cdot5n$ g/s or $0\cdot0375n$ kg/s and the heat load, $Q = 0\cdot0375n \times 2\cdot0(325 - 300) = 1\cdot875n$ kW.

Temperature driving force

$$\theta_1 = (361 - 300) = 61 \text{ K}, \quad \theta_2 = (372 - 325) = 47 \text{ K}$$

and

$$\theta_m = (61 - 47)/\ln(61/47) = 53\cdot7 \text{ K}$$

In equation 7.140:

$$X = (\theta_2 - \theta_1)/(T_1 - \theta_1) \quad \text{and} \quad Y = (T_1 - T_2)/(\theta_2 - \theta_1)$$

where T_1 and T_2 are the inlet and outlet temperatures on the shell side and θ_1 and θ_2 are the inlet and outlet temperatures on the tube side.

$$\therefore \qquad\qquad X = (325 - 300)/(372 - 300) = 0.347$$

and $\qquad\qquad\qquad Y = (372 - 361)/(325 - 300) = 0.44$

For one shell side pass, two tube side passes, Fig. 7.51 applies and $F = 0.98$.

Area

In equation 7.140, $A = Q/UF\theta_m = 1.875n/(0.230 \times 0.98 \times 53.7)$
$$= 0.155n \text{ m}^2$$

The area per unit length based on 10 mm i.d. $= (\pi \times 0.010 \times 1.0) = 0.0314 \text{ m}^2/\text{m}$ and total length of tubing $= 0.155n/0.0314 = 4.94n$ m.

Thus the length of tubes required $= (4.94n/n) = \underline{\underline{4.94 \text{ m}}}$.

Problem 7.19

A condenser consists of a number of metal pipes of outer diameter 25 mm and thickness 2.5 mm. Water, flowing at 0.6 m/s, enters the pipes at 290 K, and it is not permissible that it should be discharged at a temperature in excess of 310 K.

If 1.25 kg/s of a hydrocarbon vapour is to be condensed at 345 K on the outside of the pipes, how long should each pipe be and how many pipes would be needed?

Take the coefficient of heat transfer on the water side as 2.5, and on the vapour side as 0.8 kW/m² K and assume that the overall coefficient of heat transfer from vapour to water, based upon these figures, is reduced 20% by the effects of the pipe walls, dirt, and scale.

The latent heat of the hydrocarbon vapour at 345 K is 315 kJ/kg.

Solution

Heat load

For condensing the organic at 345 K, $Q = (1.25 \times 315) = 393.8$ kW.

If the water outlet temperature is limited to 310 K, then the mass flow of water is given by:

$$393.8 = m \times 4.18(310 - 290) \quad \text{or} \quad m = 4.71 \text{ kg/s}$$

Temperature driving force

$$\theta_1 = (345 - 290) = 55 \text{ K}, \quad \theta_2 = (345 - 310) = 35 \text{ K}$$

Therefore in equation 7.10, $\theta_m = (55 - 35)/\ln(55/35) = 44 \cdot 3$ K.
No correction factor is necessary with isothermal conditions in the shell.

Overall coefficient

Inside: $h_i = 2 \cdot 5$ kW/m^2 K.

The outside diameter $= 0 \cdot 025$ m and $d_i = (25 - 2 \times 2 \cdot 5)/10^3 = 0 \cdot 020$ m.

Basing the inside coefficient on the outer diameter,

$$h_{io} = (2 \cdot 5 \times 0 \cdot 020/0 \cdot 025) = 2 \cdot 0 \text{ kW/m}^2 \text{ K}$$

Outside: $h_o = 0 \cdot 8$ kW/m^2 K
and hence the clean overall coefficient is given by:

$$1/U_c = 1/h_{io} + 1/h_o = 1 \cdot 75 \text{ m}^2 \text{ K/kW}$$

and $$U_c = 0 \cdot 572 \text{ kW/m}^2 \text{ K}$$

Thus allowing for scale and the wall,

$$U_D = 0 \cdot 572(100 - 20)/100 = 0 \cdot 457 \text{ kW/m}^2 \text{ K}$$

Area

In equation 7.9, $A = Q/U\theta_m$
$$= 393 \cdot 8/(0 \cdot 457 \times 44 \cdot 3) = 19 \cdot 45 \text{ m}^2$$

Outside area $= (\pi \times 0 \cdot 025 \times 1 \cdot 0) = 0 \cdot 0785$ m^2/m
and hence total length of piping $= (19 \cdot 45/0 \cdot 0785) = 247 \cdot 6$ m.

$4 \cdot 71$ kg/s water $\equiv (4 \cdot 71/100) = 0 \cdot 00471$ m^3/s
and hence cross-sectional area/pass to give a velocity of $0 \cdot 6$ m/s $= (0 \cdot 00471/0 \cdot 6) = 0 \cdot 00785$ m^2.

Cross area of one tube $= (\pi/4)(0 \cdot 020)^2 = 0 \cdot 000314$ m^2.

Therefore number of tubes/pass $= (0 \cdot 00785/0 \cdot 000314) = 25$.

Thus:

with 1 tube pass, total tubes $= 25$ and tube length $= (247 \cdot 6/25) = 9 \cdot 90$ m

with 2 tube passes, total tubes $= 50$ and tube length $= (247 \cdot 6/50) = 4 \cdot 95$ m

with 4 tube passes, total tubes $= 100$ and tube length $= (247 \cdot 6/100) = 2 \cdot 48$ m

This last proposition is the most realistic.

Problem 7.20

An organic vapour is being condensed at 350 K on the outside of a nest of pipes through which water flows at $0 \cdot 6$ m/s, its inlet temperature being 290 K. The outer and inner diameters of the pipes are 19 mm and 15 mm respectively, but a layer of scale $0 \cdot 25$ mm thick and thermal conductivity $2 \cdot 0$ W/m K, has formed on the inside of the pipes.

If the coefficients of heat transfer on the vapour and water sides respectively are $1 \cdot 7$ and $3 \cdot 2$ kW/m^2 K and it is required to condense $0 \cdot 025$ kg/s of vapour on each of the pipes, how long should these be, and what will be the outlet temperature of water?

The latent heat of condensation is 330 kJ/kg.
Neglect any small resistance to heat transfer in the pipe walls.

Solution

For a total of n pipes, mass flow of vapour condensed $= 25n \times 10^{-3}$ kg/s
and hence load, $Q = 0.025n \times 330 = 8.25n$ kW.
For a water outlet temperature of T K and a mass flow of m kg/s:

$$8.25n = m \times 4.18(T - 290) \text{ kW}$$

or

$$m = 1.974n/(T - 290) \text{ kg/s} \tag{i}$$

$$\theta_1 = (350 - 290), \quad \theta_2 = (350 - T)$$

and hence in equation 7.10:

$$\theta_m = [(350 - 290) - (350 - T)]/\ln[(350 - 290)/(350 - T)]$$

$$= (T - 290)/\ln[60/(350 - T)] \quad \text{K}$$

Considering the film coefficients, $h_i = 3.2$ kW/m² K, $h_o = 1.7$ kW/m² K, and hence
$h_{io} = 3.2 \times 0.015/0.019 = 2.526$ kW/m² K.
The scale resistance,

$$(x/k) = (0.25 \times 10^{-3})/2.0 = 0.000125 \text{ m}^2 \text{ K/W} \quad \text{or} \quad 0.125 \text{ m}^2 \text{ K/kW}$$

Therefore the overall coefficient, neglecting the wall resistance is given by:

$$1/U = 1/h_{io} + x/k + 1/h_o$$

$$= 0.5882 + 0.125 + 0.396 = 1.109 \text{ m}^2 \text{ K/kW}$$

and

$$U = 0.902 \text{ kW/m}^2 \text{ K}$$

Therefore in equation 7.9:

$$A = Q/U\theta_m$$

$$= 8.25n/\{0.902(T - 290)/\ln[60/(350 - T)]\} \quad \text{m}^2$$

$$= \frac{4.18m(T - 290)\ln[60/(350 - T)]}{0.902(T - 290)}$$

$$= 4.634m \ln[60/(350 - T)] \quad \text{m}^2 \tag{ii}$$

The cross area for flow $= (\pi/4)(0.015)^2 = 0.000177$ m²/tube.
m kg/s $\equiv 0.001m$ m³/s
and area/pass to give a velocity of 0.6 m/s $= (0.001m/0.6) = 0.00167m$ m².

∴   number of tubes/pass $= (0.00167m/0.000177) = 9.42m \tag{iii}$

Area per unit length of tube $= (\pi \times 0.019 \times 1.0) = 0.0597$ m²/m.

∴   total length of tubes $= 4.634m \ln[60/(350 - T)]/0.0597$

$$= 77.6m \ln[60/(350 - T)] \text{ m}$$

length of each tube $= 77.6m \ln[60/(350 - T)]/n$ m

and, substituting from (i),

$$\text{tube length} = 77{\cdot}6 \times 1{\cdot}974n \ \ln[60/(350 - T)]/[n\,(T - 290)]$$
$$= 153{\cdot}3 \ln[60/(350 - T)]/(T - 290) \text{ m} \qquad \text{(iv)}$$

The procedure is now to select the number of tube passes N and hence m in terms of n from (iii): T is then obtained from (i) and hence the tube length from (iv). In this way the following results are obtained:

No. of tube passes N	Total tubes n	Outlet water temperature T (K)	Tube length (m)
1	9·42m	308·6	3·05
2	18·84m	327·2	3·99
4	37·68m	364·4	—
6	56·52m	401·6	—

Arrangements with 4 and 6 tube side passes require water outlet temperatures in excess of the condensing temperature and are clearly not possible. With 2 tube side passes, $T = 327{\cdot}2$ K at which severe scaling is promoted and hence the proposed unit would consist of one tube side pass and a tube length of 3·05 m.

The outlet water temperature would be 308·6 K.

Problem 7.21

A heat exchanger is required to cool continuously 20 kg/s of warm water from 360 K to 335 K by means of 25 kg/s of cold water, inlet temperature 300 K. Assuming that the water velocities are such as to give an overall coefficient of heat transfer of 2 kW/m² K, assumed constant, calculate the total area of surface required (a) in a counterflow heat exchanger, i.e. one in which the hot and cold fluids flow in opposite directions, and (b) in a multi-pass heat exchanger, with the cold water making two passes through the tubes, and the hot water making one pass along the outside of the tubes. In case (b) assume that the hot-water flows in the same direction as the inlet cold water, and that its temperature over any cross-section is uniform.

Solution

The heat load, $Q = 20 \times 4{\cdot}18(360 - 335) = 2090$ kW
and the outlet cold water temperature is given by:

$$2090 = 25 \times 4{\cdot}18(T_2 - 300) \quad \text{or} \quad T_2 = 320 \text{ K}$$

Case (a)

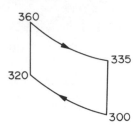

$\theta_1 = (360 - 320) = 40 \text{ K}, \quad \theta_2 = (335 - 300) = 35 \text{ K}$

and from equation 7.10:

$$\theta_m = (40 - 35)/\ln(40/35) = 37\cdot4 \text{ K}$$

As the flow is true counter flow, no correction factor is necessary and $F = 1\cdot0$. Therefore in equation 7.140:

$$A = Q/UF\theta_m = 2090/(2\cdot0 \times 1\cdot0 \times 37\cdot4)$$
$$= 27\cdot94 \text{ m}^2$$

Case (b)

Again, $\theta_m = 37\cdot4 \text{ K}$

$$X = (\theta_2 - \theta_1)/(T_1 - \theta_1) = (320 - 300)/(360 - 300) = 0\cdot33$$
$$Y = (T_1 - T_2)/(\theta_2 - \theta_1) = (360 - 335)/(320 - 300) = 1\cdot25$$

Hence, from figure 7.51, $F = 0\cdot94$
and in equation 7.140:

$$A = 2090/(2\cdot0 \times 0\cdot94 \times 374)$$
$$= 29\cdot73 \text{ m}^2$$

Problem 7.22

Find the heat loss per m^2 of surface through a brick wall 0·5 m thick when the inner surface is at 400 K and the outside at 310 K: the thermal conductivity of the brick may be taken as 0·7 W/m K.

Solution

In equation 7.11: $Q = kA(T_1 - T_2)/x$
$$= 0\cdot7 \times 1\cdot0(400 - 310)/0\cdot5 = 126 \text{ W/m}^2$$

Problem 7.23

A furnace is constructed with 225 mm of firebrick, 120 mm of insulating brick, and 225 mm of building brick. The inside temperature is 1200 K and the outside temperature 330 K. If the thermal conductivities are 1·4, 0·2, and 0·7 W/m K, find the heat loss per unit area and the temperature at the junction of the firebrick and insulating brick.

Solution

Let T_1 K and T_2 K be the temperatures at the firebrick/insulating brick and the insulating brick/building brick junctions respectively.
Using equation 7.11:
for conduction through the firebrick:

$$Q = 1 \cdot 4 \times 1 \cdot 0(1200 - T_1)/0 \cdot 225$$
$$= 6 \cdot 22(1200 - T_1) \text{ W/m}^2 \tag{i}$$

for conduction through the insulating brick:

$$Q = 0 \cdot 2 \times 1 \cdot 0(T_1 - T_2)/0 \cdot 120$$
$$= 1 \cdot 67(T_1 - T_2) \text{ W/m}^2 \tag{ii}$$

for conduction through the building brick:

$$Q = 0 \cdot 7 \times 1 \cdot 0(T_2 - 330)/0 \cdot 225$$
$$= 3 \cdot 11(T_2 - 330) \text{ W/m}^2 \tag{iii}$$

Thus the thermal resistances of each material (x/kA) are: firebrick $= (1/6 \cdot 22) = 0 \cdot 161$; insulating brick $= (1/1 \cdot 67) = 0 \cdot 60$; building brick $= (1/3 \cdot 11) = 0 \cdot 322$ K/Wm^2; and in equation 7.14:
$$(1200 - 330) = (0 \cdot 161 + 0 \cdot 60 + 0 \cdot 322)Q$$
or
$$\underline{\underline{Q = 803 \cdot 3 \text{ W/m}^2}}$$

Now, ΔT firebrick$/\Sigma \Delta T = (x/kA)_{\text{firebrick}}/\Sigma(x/kA)$

$\therefore$ $(1200 - t_1)/(1200 - 330) = 0 \cdot 161/(0 \cdot 161 + 0 \cdot 60 + 0 \cdot 322) = 0 \cdot 161/1 \cdot 083$

and
$$\underline{\underline{T_1 = 1071 \text{ K}}}$$

Similarly for the insulating brick:

$$(1071 - T_2)/(1200 - 330) = 0 \cdot 60/1 \cdot 083$$

and
$$\underline{\underline{T_2 = 589 \text{ K}}}$$

Problem 7.24

Calculate the total heat loss by radiation and convection from an unlagged horizontal steam pipe of 50 mm outside diameter at 415 K to air at 290 K.

Solution

Outside area per unit length of pipe $= (\pi \times 0 \cdot 050 \times 1 \cdot 0) = 0 \cdot 157$ m^2/m.

Convection

For natural convection from a horizontal pipe to air, the simplified form of equation 7.78 may be used:

$$h_c = 1 \cdot 18 (\Delta T/d_o)^{0 \cdot 25}$$

In this case, $\Delta T = (415 - 290) = 125$ K and $d_o = 0 \cdot 050$ m.

$$\therefore \qquad\qquad h_c = 1 \cdot 18 (125/0 \cdot 050)^{0 \cdot 25} = 8 \cdot 34 \ \text{W/m}^2 \ \text{K}$$

Thus, heat loss by convection, $\qquad q_c = h_c A (T_1 - T_2)$

$$= 8 \cdot 34 \times 0 \cdot 157 (415 - 290) = \underline{\underline{163 \cdot 7 \ \text{W/m}}}$$

Radiation

Equation 7.84 may be used. Taking the emissivity of the pipe as $0 \cdot 9$:

$$q_r = 5 \cdot 67 \times 10^{-8} \times 0 \cdot 9 (415^4 - 290^4) \times 0 \cdot 157 = \underline{\underline{181 \cdot 0 \ \text{W/m}}}$$

and the total loss is $\underline{\underline{344 \cdot 7 \ \text{W/m}}}$ length of pipe.

Problem 7.25

Toluene is continuously nitrated to mononitrotoluene in a cast-iron vessel of 1 m diameter fitted with a propeller agitator of $0 \cdot 3$ m diameter driven at 2 Hz. The temperature is maintained at 310 K by circulating cooling water at $0 \cdot 5$ kg/s through a stainless steel coil of 25 mm outside diameter and 22 mm inside diameter wound in the form of a helix of $0 \cdot 81$ m diameter. The conditions are such that the reacting material may be considered to have the same physical properties as 75% sulphuric acid. If the mean water temperature is 290 K, what is the overall heat transfer coefficient?

Solution

The overall coefficient U_0 based on the outside area of the coil is given by equation 7.130:

$$1/U_o = 1/h_o + (x_w/k_w)(d_o/d_w) + (1/h_i)(d_o/d_i) + R_o + R_i(d_o/d_i)$$

where d_w is the mean pipe diameter.

Inside

The coefficient on the water side is given by equations 7.131 and 7.132:

$$h_i = (k/d)(1 + 3 \cdot 5d/d_c)0 \cdot 023(d_i u \rho/\mu)^{0 \cdot 8}(C_p \mu/k)^{0 \cdot 4}$$

where $\qquad u \rho = 0 \cdot 5/[(\pi/4) \times 0 \cdot 022^2] = 1315 \ \text{kg/m}^2 \ \text{s}$

$$d_i = 0 \cdot 022 \text{ m}, \quad d_c = 0 \cdot 80 \text{ m}$$

and for water at 290 K: $k = 0 \cdot 59$ W/m K, $\mu = 0 \cdot 00108$ N s/m^2, and $C_p = 4180$ J/kg K.

$\therefore$
$$h_i = (0 \cdot 59/0 \cdot 022)(1 + 3 \cdot 5 \times 0 \cdot 022/0 \cdot 80) \times 0 \cdot 023(0 \cdot 022$$
$$\times 1315/0 \cdot 00108)^{0 \cdot 8}(4180 \times 0 \cdot 00108/0 \cdot 59)^{0 \cdot 4}$$
$$= 0 \cdot 680(26,780)^{0 \cdot 8}(7 \cdot 65)^{0 \cdot 4} = 5490 \text{ W/m}^2 \text{ K}$$

Outside

In equation 7.133:

$$(h_o d_v/k)(\mu_s/\mu)^{0 \cdot 14} = 0 \cdot 87(C_p \mu/k)^{0 \cdot 33}(L^2 N \rho/\mu)^{0 \cdot 62}$$

For 75% sulphuric acid:

$k = 0 \cdot 40$ W/m K

$\mu_s = 0 \cdot 0086$ N s/m^2 at 300 K, $\quad \mu = 0 \cdot 0065$ N s/m^2 at 310 K

$C_p = 1880$ J/kg K

$\rho = 1666$ kg/m^3

$\therefore$
$$(h_o \times 1 \cdot 0/0 \cdot 40)(0 \cdot 0086/0 \cdot 0065)^{0 \cdot 14} = 0 \cdot 87(1880 \times 0 \cdot 0065/0 \cdot 40)^{0 \cdot 33}$$
$$\times (0 \cdot 3^2 \times 2 \cdot 0 \times 1665/0 \cdot 0065)^{0 \cdot 62}$$

$\therefore$
$$2 \cdot 5 h_o(1 \cdot 323)^{0 \cdot 14} = 0 \cdot 87(30 \cdot 55)^{0 \cdot 33}(46,108)^{0 \cdot 62}$$

and
$$h_o = 0 \cdot 348 \times 3 \cdot 09 \times 779/1 \cdot 04 = 805 \cdot 5 \text{ W/m}^2 \text{ K}$$

Overall

Taking $k_w = 15 \cdot 9$ W/m K and R_o and R_i as $0 \cdot 0004$ and $0 \cdot 0002$ m^2 K/W, then in equation 7.130:

$$1/U_o = (1/805 \cdot 5) + (0 \cdot 0015/15 \cdot 9)(0 \cdot 025/0 \cdot 0235) + (1/5490)(0 \cdot 025/0 \cdot 022)$$
$$+ 0 \cdot 0004 + 0 \cdot 0002(0 \cdot 025/0 \cdot 022)$$
$$= (0 \cdot 00124 + 0 \cdot 00010 + 0 \cdot 00021 + 0 \cdot 00040 + 0 \cdot 00023) = 0 \cdot 00218 \text{ m}^2 \text{ K/W}$$

and
$$\underline{\underline{U_o = 458 \cdot 7 \text{ W/m}^2 \text{ K}}}$$

Problem 7.26

7·5 kg/s of pure iso-butane are to be condensed at a temperature of 331·7 K in a horizontal tubular exchanger using a water inlet temperature of 301 K. It is proposed to use 19 mm outside diameter tubes of 1·6 mm wall arranged on a 25 mm triangular pitch. Under these conditions the resistance of the scale may be taken as 0·0005 m^2 K/W. It is required to determine the number and arrangement of the tubes in the shell.

Solution

The latent heat of vaporisation of isobutane is 286 kJ/kg and hence the heat load,

$$Q = (7 \cdot 5 \times 286) = 2145 \text{ kW}$$

The cooling water outlet should not exceed 320 K and a value of 315 K will be used. The mass flow of water is then

$$2145/[4 \cdot 18(315 - 301)] = 36 \cdot 7 \text{ kg/s}$$

In order to obtain an approximate size of the unit, a value of 500 W/m² K will be assumed for the overall coefficient based on the outside area of the tubes.

$$\theta_1 = (331 \cdot 7 - 301) = 30 \cdot 7 \text{ K}, \quad \theta_2 = (331 \cdot 7 - 315) = 16 \cdot 7 \text{ K}$$

and from equation 7.10: $\qquad \theta_m = (30 \cdot 7 - 16 \cdot 7)/\ln(30 \cdot 7/16 \cdot 7) = 23 \cdot 0 \text{ K}.$

Thus, the approximate area $= 2145 \times 10^3/(500 \times 23 \cdot 0) = 186 \cdot 5 \text{ m}^2$.

The outside area of 0·019 m diameter tubes $= (\pi \times 0 \cdot 019 \times 1 \cdot 0) = 0 \cdot 0597 \text{ m}^2/\text{m}$ and hence the total length of tubing $= (186 \cdot 5/0 \cdot 0597) = 3125 \text{ m}$.

Adopting a standard tube length of 4·88 m, number of tubes $= (3125/4 \cdot 88) = 640$. With the large flow of water involved, a four-tube side pass unit is proposed, and for this arrangement 678 tubes can be accommodated on 25 mm triangular pitch in a 0·78 m i.d. shell. Using this layout, the film coefficients will now be estimated and hence the assumed value of U checked.

Inside

Water flow through each tube $= 36 \cdot 7/(678/4) = 0 \cdot 217 \text{ kg/s}$.

The tube i.d. $= (19 \cdot 0 - 2 \times 1 \cdot 67) = 15 \cdot 7 \text{ mm}$
and the cross-sectional area for flow $= (\pi/4)(0 \cdot 0157)^2 = 0 \cdot 000194 \text{ m}^2$
and hence the water velocity, $u = 0 \cdot 217/(1000 \times 0 \cdot 000194) = 1 \cdot 12 \text{ m/s}$.

From equation 7.59: $\qquad h_i = 1063(1 + 0 \cdot 00293 \times 308)1 \cdot 12^{0 \cdot 8}/0 \cdot 0157^{0 \cdot 2}$

$$= 1063 \times 1 \cdot 903 \times 1 \cdot 095/0 \cdot 436 = 5084 \text{ W/m}^2 \text{ K}$$

or, based on outside diameter,

$$h_{io} = 5084 \times 0 \cdot 0157/0 \cdot 019 = 4201 \text{ W/m}^2 \text{ K} \quad \text{or} \quad 4 \cdot 20 \text{ kW/m}^2 \text{ K}$$

Outside

The temperature drop across the condensate film, ΔT_f is given by:

(thermal resistance of water film + scale)/(total thermal resistance) $= (\theta_m - \Delta T_f)/\theta_m$

or $\qquad\qquad\qquad (1/4 \cdot 20 + 0 \cdot 5)/(1/0 \cdot 500) = (23 \cdot 0 - \Delta T_f)/23 \cdot 0$

and $\qquad\qquad\qquad\qquad\qquad \Delta T_f = 14 \cdot 5 \text{ K}$

The condensate film is thus at $(331 \cdot 7 - 14 \cdot 5) = 317 \cdot 2 \text{ K}$.

The outside film coefficient is given by equation 7.108:

$$h_o = 0 \cdot 72[(k^3 \rho^2 g \lambda)/(j d_o \mu \Delta T_f)]^{0 \cdot 25}$$

At 317·2 K, $k = 0·13$ W/m K, $\rho = 508$ kg/m³, $\mu = 0·000136$ N s/m², and $j = \sqrt{678} = 26·0$.

$\therefore$
$$h_o = 0·72[(0·13^3 \times 508^2 \times 9·81 \times 286 \times 10^3)/(26 \times 19·0 \times 10^{-3}$$
$$\times 0·000136 \times 14·5)]^{0·25}$$
$$= 8·14 \text{ W/m}^2 \text{ K} \quad \text{or} \quad 0·814 \text{ kW/m}^2 \text{ K}$$

Overall

$$1/U = (1/4·20) + (1/0·814) + 0·50 = 1·967$$
and
$$U = 0·508 \text{ kW/m}^2 \text{ K} \quad \text{or} \quad 508 \text{ W/m}^2 \text{ K}$$

which is sufficiently near the assumed value. For the proposed unit, the heat load is

$$Q = 0·508 \times 678 \times 4·88 \times 0·0597 \times 23·0 = 2310 \text{ kW}$$

or an overload of $\qquad (2310 - 2145)100/2145 = 7·7\%$

Problem 7.27

37·5 kg/s of crude oil is to be heated from 295 to 330 K by heat exchange with the bottom product from a distillation column. The bottom product, flowing at 29·6 kg/s is to be cooled from 420 to 380 K. There is available a tubular exchanger with an inside shell diameter of 0·60 m, having one pass on the shell side and two passes on the tube side. It has 324 tubes, 19 mm outside diameter with 2·1 mm wall and 3·65 m long, arranged on a 25 mm square pitch and supported by baffles with a 25% cut, spaced at 230 mm intervals. Would this exchanger be suitable?

Solution

Mean temperature of bottom product $= 0·5(420 + 380) = 400$ K.
Mean temperature of crude oil $= 0·5(330 + 295) = 313$ K.
For the crude oil at 313 K; $C_p = 1·986 \times 10^3$ J/kg K.
$$\mu = 0·0029 \text{ N s/m}^2.$$
$$k = 0·136 \text{ W/m K}.$$
$$\rho = 824 \text{ kg/m}^3.$$
For the bottom product at 400 K: $C_p = 2200$ J/kg K.
Heat loads:

tube side: $Q = 37·5 \times 1·986(330 - 295) = 2607$ kW.
shell side: $Q = 29·6 \times 2·20(420 - 380) = 2605$ kW.

Outside coefficient

Temperature of wall $= 0·5(400 + 313) = 356·5$ K

and film temperature, $T_f = 0.5(400 + 356.5) = 378$ K.

At 378 K, $\rho = 867$ kg/m³, $\mu = 0.0052$ N s/m², and $k = 0.119$ W/m K.

Cross area for flow = (shell i.d. × clearance × baffle spacing)/pitch
$$= (0.60 \times 0.0064 \times 0.23)/0.025 = 0.0353 \text{ m}^2$$

(assuming 0·0064 m clearance).

∴ $G'_{max} = (29.6/0.0353) = 838.5$ kg/m² s

and $Re_{max} = (0.019 \times 838.5/0.0052) = 306.4$

Therefore in equation 7.68, taking $C_h = 1$:

$$h_o \times 0.019/0.119 = 0.33 \times 1.0(3064)^{0.6}(2200 \times 0.0052/0.119)^{0.3}$$

or $h_o = (2.07 \times 125 \times 3.94) = 1018$ W/m² K
$$\equiv 1.02 \text{ kW/m}^2 \text{ K}$$

Inside coefficient

Number tubes per pass = 324/2 = 162.

Inside diameter = 0·0148 m

and cross-sectional area for flow = $(\pi/4)(0.0148)^2 = 0.000172$ m² per tube or $(0.000172 \times 162) = 0.0279$ m² per pass.

∴ $G' = (37.5/0.0279) = 1346$ kg/m² s

In equation 7.50:

$$h_i \times 0.0148/0.136 = 0.023(0.0148 \times 1346/0.0029)^{0.8}(1986 \times 0.0029/0.136)^{0.4}$$
$$h_i = 0.211(6869)^{0.8}(42.4)^{0.4}$$
$$= 1110 \text{ W/m}^2 \text{ K}$$

or, based on the outside area, $h_{io} = 1110 \times 0.0148/0.019 = 865$ W/m² K or 0·865 kW/m² K.

Overall coefficient

Neglecting the wall and scale resistance, the clean overall coefficient is given by:

$$1/U_c = (1/1.02) + (1/0.865) = 2.136 \text{ K m}^2/\text{kW}$$

The area available is $A = (324 \times 3.65 \times \pi \times 0.019) = 70.7$ m² and hence the minimum value of the design coefficient is given by

$$1/U_D = A\theta_m/Q$$
$$\theta_1 = (420 - 330) = 90 \text{ K}, \quad \theta_2 = (380 - 295) = 85 \text{ K}$$

and $\theta_m = (90 - 85)/\ln(90/85) = 87.5$ K

∴ $1/U_D = (70.7 \times 87.5/2607) = 2.37$ m² K/kW

The maximum allowable scale resistance is then:

$$R = (1/U_D) - (1/U_c) = (2 \cdot 37 - 2 \cdot 136) = 0 \cdot 234 \text{ m}^2 \text{ K/kW}$$

This value is very low as seen from Table 7.11, and the exchanger would not give the required temperatures without frequent cleaning.

Problem 7.28

A 150 mm internal diameter steam pipe is carrying steam at 444 K and is lagged with 50 mm of 85% magnesia. What will be the heat loss to the air at 294 K?

Solution

In this case, $d_i = 0 \cdot 150$ m

$$d_o = 0 \cdot 168 \ m \text{ and } d_w = 0 \cdot 5(0 \cdot 150 + 0 \cdot 168) = 0 \cdot 159 \text{ m}$$
$$d_s = (0 \cdot 168 + 2 \times 0 \cdot 050) = 0 \cdot 268 \text{ m}$$

and d_m, the log mean of d_o and $d_s = 0 \cdot 215$ m.

The coefficient for condensing steam plus any scale will be taken as 8500 W/m² K, k_w as 45 W/m K, and k_l as 0·073 W/m K.

The surface temperature of the lagging will be assumed as 314 K and $(h_r + h_c)$ as 10 W/m² K.

The thermal resistances are therefore:

$$(1/h_i \pi d) = 1/(8500 \times \pi \times 0 \cdot 150) = 0 \cdot 00025 \text{ mK/W}$$
$$(x_w/k_w \pi d_w) = 0 \cdot 009/(45 \pi \times 0 \cdot 159) = 0 \cdot 00040 \text{ mK/W}$$
$$(x_l/k_l \pi d_m) = 0 \cdot 050/(0 \cdot 073 \pi \times 0 \cdot 215) = 1 \cdot 0130 \text{ mK/W}$$
$$(1/(h_r + h_c)d_s) = 1/(10 \times 0 \cdot 268) = 0 \cdot 1190 \text{ mK/W}$$

Neglecting the first two terms, the total thermal resistance = 1·132 mK/W. From equation 7.154, heat lost per unit length = $(444 - 294)/1 \cdot 132 = \underline{\underline{132 \cdot 5 \text{ W/m}}}$.

The surface temperature of the lagging is then given by:

$$\Delta T (\text{lagging})/\Sigma \Delta T = (1 \cdot 013/1 \cdot 132) = 0 \cdot 895$$

and $$\Delta T (\text{lagging}) = 0 \cdot 895(444 - 294) = 134 \text{ K}$$

Therefore the surface temperature = $(444 - 134) = 310$ K which approximates to the assumed value.

Assuming an emissivity of 0·9, $h_r = 5 \cdot 67 \times 10^{-8} \times 0 \cdot 9(310^4 - 294^4)/(310 - 294) = 5 \cdot 63$ W/m² K.

For natural convection, $h_c = 1 \cdot 37(T/d_s)^{0 \cdot 25} = 1 \cdot 37[(310 - 294)/0 \cdot 268]^{0 \cdot 25} = 3 \cdot 81$ W/m² K.

$$\therefore \qquad (h_r + h_c) = 9 \cdot 45 \text{ W/m}^2 \text{ K}$$

which agrees with the assumed value.

In practice forced convection currents are usually present and the heat loss would probably be higher than this value.

For an unlagged pipe and $\Delta T = 150\,K$, $(h_r + h_c)$ would be about $20\,W/m^3\,K$ and the heat loss, $Q/l = (h_r + h_c)\pi d_o \Delta T$

$$= 20\pi \times 0\cdot168 \times 150 = 1584\,W/m$$

Thus the heat loss has been reduced by about 90% by the addition of 50 mm of lagging.

Problem 7.29

A refractory material which has an emissivity of 0·40 at 1500 K and 0·43 at 1420 K is at a temperature of 1420 K and is exposed to black furnace walls at a temperature of 1500 K. What is the rate of gain of heat by radiation per unit area?

Solution

In the absence of further data, the system will be considered as two parallel plates.

The radiating source is the furnace walls at $T_1 = 1500\,K$ and for a black surface, $e_1 = 1\cdot0$.

The heat sink is the refractory at $T_2 = 1420\,K$, at which $e_2 = 0\cdot43$.

From equation 7.89:

$$q = (e_1 e_2 \sigma)(T_1{}^4 - T_2{}^4)/(e_1 + e_2 - e_1 e_2)$$
$$= (1\cdot0 \times 0\cdot43 \times 5\cdot67 \times 10^{-8})(1500^4 - 1420^4)/(1\cdot0 + 0\cdot43 - 0\cdot43 \times 1\cdot0)$$
$$= 2\cdot44 \times 10^{-8}(9\cdot97 \times 10^{11})/1\cdot0$$
$$= 2\cdot43 \times 10^4\,W/m^2 \quad \text{or} \quad \underline{\underline{24\cdot3\,kW/m^2}}$$

Problem 7.30

The total emissivity of clean chromium as a function of surface temperature, T K, is given approximately by the empirical expression:

$$e = 0\cdot38(1 - 263/T)$$

Obtain an expression for the absorptivity of solar radiation as a function of surface temperature, and compare the absorptivity and emissivity at 300, 400 and 1000 K.

Assume that the sun behaves as a black body at 5500 K.

Solution

For a black body at 5500 K, the intensity of solar radiation:

$$I_s = 5 \cdot 67 \times 10^{-8}(5500)^4$$
$$= 5 \cdot 188 \times 10^7 \, \text{W/m}^2 \quad \text{sun surface}$$

Assuming radius of sun $= 7 \times 10^5$ km and the distance sun to earth $= 150 \times 10^6$ km.
$\therefore$ surface of sun $= 4\pi r_s^2 = 4\pi \times (7 \cdot 00^2 \times 10^5)^2$ km².
surface area of sphere of radius 150×10^6 km, i.e. at earth's surface, $= 4\pi(150 \times 10^6)^2$.
$\therefore$ ratio of projected areas, $A_1/A_2 = (150 \times 10^6)^2/(7 \times 10^5)^2 = 4 \cdot 59 \times 10^4$
and intensity of solar radiation at earth's surface $= 5 \cdot 188 \times 10^7/(4 \cdot 59 \times 10^4)$
$$= 1 \cdot 13 \times 10^3 \, \text{W/m}^2$$

At equilibrium, the fraction of solar radiation absorbed = radiation emitted from the plate, or

$$a_s I_s = \epsilon \sigma T^4$$
$$a_s 1 \cdot 13 \times 10^3 = 0 \cdot 38(1 - 263/T)5 \cdot 67 \times 10^{-8} T^4$$

and
$$a_s = 1 \cdot 91 \times 10^{-11}(1 - 263/T)T^4$$
$$= (1 \cdot 91 \times 10^{-11} T^4 - 5 \cdot 02 \times 10^{-9} T^3) \qquad \text{(i)}$$

At 300 K: $\quad e = 0 \cdot 38(1 - 263/300) = 0 \cdot 0467$
$$a_s = 1 \cdot 91 \times 10^{-11} \times 300^4 - 5 \cdot 02 \times 10^{-9} \times 300^3 = 0 \cdot 0192$$

At 400 K; $\quad e = 0 \cdot 38(1 - 263/400) = 0 \cdot 130$
$$a_s = 1 \cdot 91 \times 10^{-11} \times 400^4 - 5 \cdot 02 \times 10^{-9} \times 400^3 = 0 \cdot 167$$

At 1000 K: $\quad e = 0 \cdot 38(1 - 263/1000) = 0 \cdot 280$
$$a_s = 1 \cdot 91 \times 10^{-11} \times 1000^4 - 5 \cdot 02 \times 10^{-9} \times 1000^3 = 14 \cdot 08$$

Clearly this is an impossible situation as the plate will never attain 1000 K due to solar radiation alone. In fact, putting equation (i) equal to the heat radiated from the plate ($= \epsilon \sigma T^4$), an equilibrium temperature of 550 K is attained.

Problem 7.31

Repeat Problem 7.30 for the case of aluminium, assuming the emissivity to be $1 \cdot 25e$.

Solution

In this case, from Problem 7.30:

$$a_s \times 1 \cdot 13 \times 10^3 = 1 \cdot 25 \times 0 \cdot 38(1 - 263/T)5 \cdot 67 \times 10^{-8} T^4$$

and
$$a_s = (2 \cdot 38 \times 10^{-11} T^4 - 6 \cdot 27 \times 10^{-9} T^3) \qquad \text{(i)}$$

At 300 K: $e = 0.475(1 - 263/300) = 0.0584$

$a_s = (2.38 \times 10^{-11} \times 300^4 - 6.27 \times 10^{-9} \times 300^3) = 0.0235$

At 400 K: $e = 0.475(1 - 263/400) = 0.130$

$a_s = (2.38 \times 10^{-11} \times 400^4 - 6.27 \times 10^{-9} \times 400^3) = 0.208$

At 1000 K: $e = 0.475(1 - 263/1000) = 0.350$

$a_s = (2.38 \times 10^{-11} \times 1000^4 - 6.27 \times 10^{-9} \times 1000^3) = 17.6$

At equilibrium, assuming the heat absorbed is all radiated, then a surface temperature of 535 K is obtained.

Problem 7.32

Calculate the heating due to solar radiation on the flat concrete roof of a building, 8 m by 9 m, in Africa, if the surface temperature of the roof is 330 K. What would be the effect of covering the roof with a highly reflecting surface, such as polished aluminium?

The total emissivity of concrete at 330 K is 0.89, whilst the total absorptivity of solar radiation (sun temperature = 5500 K) at this temperature is 0.60.

Use the data of Problem 7.31 for aluminium.

Solution

From Problem 7.30, taking the sun as a black body at 5500 K, the intensity of solar radiation, $I_s = 1.13 \times 10^3$ W/m^2.

Area of roof = $(8 \times 9) = 72$ m^2 and $a_s = 0.60$.

Therefore heat received by the concrete = $(1.13 \times 10^3 \times 72 \times 0.60) = 4.88 \times 10^4$ W.

Heat emitted by the surface = $\epsilon \sigma A T^4$

$= 0.89 \times 5.67 \times 10^{-8} \times 72 \times 330^4 = 4.309 \times 10^4$ W

and hence the net heat absorbed by the concrete = $(4.88 - 4.309) \times 10^4$

$= 5.71 \times 10^3$ W or 5.71 kW

From Problem 7.31, the absorptivity of solar radiation for aluminium:

$$a_s = (2.38 \times 10^{-11} T^4 - 6.27 \times 10^{-9} T^3)$$

or at $T = 330$ K:

$$a_s = (2.38 \times 10^{-11} \times 330^4 - 6.27 \times 19^{-9} \times 330^3) = 0.057$$

Therefore heat received by the concrete = $(1.13 \times 10^3 \times 72 \times 0.057) = 4.638 \times 10^3$ W.

The heat emitted by the surface

$$= 1.25 \times 0.38(1 - 263/330) \times 5.67 \times 10^{-8} \times 72 \times 330^4$$
$$= 4.675 \times 10^3 \text{ W}$$

and hence the net heat lost by the aluminium $= (4\cdot675 - 4\cdot638) \times 10^3$
$$= 0\cdot037 \times 10^3 \text{ W or } \underline{\underline{0\cdot037 \text{ kW}}}$$

Problem 7.33

A rectangular iron ingot $15 \text{ cm} \times 15 \text{ cm} \times 30 \text{ cm}$ is supported at the centre of a reheating furnace. The furnace has walls of silica-brick at 1400 K, and the initial temperature of the ingot is 290 K. How long will it take to heat the ingot to 600 K?

It may be assumed that the furnace is large compared with the ingot-size, and that the ingot remains at uniform temperature throughout its volume; convection effects are negligible.

The total emissivity of the oxidised iron surface is 0·78 and both emissivity and absorptivity are independent of the surface temperature.

Density of iron $= 7\cdot2 \text{ Mg/m}^3$; specific heat of iron $= 0\cdot50 \text{ kJ/kg K}$

Solution

As there are no temperature gradients within the ingot, the rate of heating is dependent on the rate of radiative heat transfer to the surface. In addition, since the dimensions of the ingot are much smaller than those of the surrounding surfaces, these may be treated as black.

Volume of ingot $= (15 \times 15 \times 30) = 6750 \text{ cm}^3$ or $0\cdot00675 \text{ m}^3$.

Mass of ingot $= (7\cdot2 \times 10^3 \times 0\cdot00675) = 48\cdot6 \text{ kg}$.

For an ingot temperature of T K, the increase in enthalpy $= \mathrm{d}(mC_pT)/\mathrm{d}t$ or $mC_p \, \mathrm{d}T/\mathrm{d}t$

where t s is the time and C_p the specific heat of the ingot.

The heat received by radiation $= A\sigma a(T_f^4 - T^4)$

where the area, $A = (4 \times 30 \times 15) + (2 \times 15 \times 15) = 2250 \text{ cm}^2$ or $0\cdot225 \text{ m}^2$,

the absorptivity a will be taken as the emissivity $= 0\cdot78$

and the furnace temperature, $T_f = 1400 \text{ K}$.

$$\therefore \ mC_p \, \mathrm{d}T/\mathrm{d}t = A\sigma a(T_f^4 - T^4)$$

or
$$\int_0^t \mathrm{d}t = \frac{mC_p}{aA\sigma} \int_{290}^{600} \frac{\mathrm{d}T}{(T_f^4 - T^4)}$$

$$\therefore$$
$$t = \frac{(48\cdot6 \times 0\cdot50 \times 10^3)}{(0\cdot78 \times 0\cdot225 \times 5\cdot67 \times 10^{-8})(4 \times 1400^3)} \left(\ln \frac{1400+T}{1400-T} + 2 \tan^{-1} \frac{T}{1400} \right)_{290}^{600}$$

$$= \underline{\underline{200 \text{ s}}}$$

Problem 7.34

A wall is made of brick, of thermal conductivity 1·0 W/m K, 230 mm thick, lined on the inner face with plaster of thermal conductivity 0·4 W/m K and of thickness

10 mm. If a temperature difference of 30 K is maintained between the two outer faces, what is the heat flow per unit area of wall?

Solution

For an area of 1 m²:

thermal resistance of the brick, $(x_1/k_1A) = 0.230/(1.0 \times 1.0) = 0.230$ K/W
thermal resistance of the plaster, $(x_2/k_2A) = 0.010/(0.4 \times 1.0) = 0.0025$ K/W

and in equation 7.14: $30 = (0.230 + 0.0025)Q$

or $Q = 129$ W

Problem 7.35

A 50 mm diameter pipe of circular cross-section and with walls 3 mm thick is covered with two concentric layers of lagging, the inner layer having a thickness of 25 mm and a thermal conductivity of 0.08 W/m K, and the outer layer has a thickness of 40 mm and a thermal conductivity of 0.04 W/m K. What is the rate of heat loss per metre length of pipe if the temperature inside the pipe is 550 K and the outside surface temperature is 330 K?

Solution

From equation 7.18, the thermal resistance of each component is:
$$(r_2 - r_1)/k(2\pi r_m l)$$
Thus *for the wall*, $r_2 = (0.050/2) + 0.003 = 0.028$ m
$r_1 = (0.050/2) = 0.025$ m
and $r_m = (0.028 - 0.025)/(\ln 0.028/0.025) = 0.0265$ m.
Taking $k = 45$ W/m K and $l = 1.0$ m the thermal resistance $= (0.028 - 0.025)/(45 \times 2\pi \times 0.0265 \times 1.0) = 0.00040$ K/W.
For the *inner lagging*, $r_2 = (0.028 + 0.025) = 0.053$ m
$r_1 = 0.028$ m
and $r_m = (0.053 - 0.028)/(\ln 0.053/0.028) = 0.0392$ m.
Therefore thermal resistance $= (0.053 - 0.028)/(0.08 \times 2\pi \times 0.0392 \times 1.0)$
$= 1.2688$ K/W
For the *outer lagging*: $r_2 = (0.053 + 0.040) = 0.093$ m
$r_2 = 0.053$ m
and $r_m = (0.093 - 0.053)/(\ln 0.093/0.053) = 0.0711$ m.
Therefore thermal resistance $= (0.093 - 0.053)/(0.04 \times 2\pi \times 0.0711 \times 1.0)$
$= 2.2385$ K/W
From equation 7.15: $Q = (550 - 330)/(0.0004 + 1.2688 + 2.2385)$
$= 62.7$ W/m

Problem 7.36

The temperature of oil leaving a co-current flow cooler is to be reduced from 370 to 350 K by lengthening the cooler. The oil and water flowrates and inlet temperatures, and the other dimensions of the cooler, will remain constant. The water enters at 285 K and oil at 420 K. The water leaves the original cooler at 310 K. If the original length is 1 m, what must be the new length?

Solution

For the *original cooler*: $Q = m_o s_o (420 - 370)$ for the oil

and $Q = m_w s_w (310 - 285)$ for the water

∴ $(m_o s_o / m_w s_w) = (25/50) = 0.5$

where m_o and m_w are the mass flows and s_o and s_w the specific heats of the oil and water respectively.

$\theta_1 = (420 - 285) = 135$ K, $\theta_2 = (370 - 310) = 60$ K for co-current flow,

and from equation 7.10: $\theta_m = (135 - 60)/\ln(135/60) = 92.5$ K

If a is the area per unit length of tube multiplied by the number of tubes, then $A = 1.0 \times a$ m^2 and in equation 7.9:

$$m_o s_o (420 - 370) = Ua 92.5 \quad \text{or} \quad m_o s_o / Ua = 1.85$$

For the *new cooler*: $Q = m_o s_o (420 - 350)$ for the oil

and $Q = m_w s_w (T - 285)$ for the water

where T is the water outlet temperature.

∴ $(T - 285) = (m_o s_o / m_w s_w)(420 - 350)$

$$= 0.5 \times 70$$

and $T = 320$ K

∴ $\theta_1 = (420 - 285) = 135$ K, $\theta_2 = (350 - 320) = 30$ K

and from equation 7.10: $\theta_m = (135 - 30)/\ln(135/30) = 69.8$ K

In equation 7.9: $m_o s_o (420 - 350) = Ual 69.8$

∴ $l = (m_o s_o / Ua) \times 1.003$

$$= (1.85 \times 1.003) = \underline{\underline{1.86 \text{ m}}}$$

Problem 7.37

In a countercurrent-flow heat exchanger, 1.25 kg/s of benzene (specific heat 1.9 kJ/kg K and specific gravity 0.88) is to be cooled from 350 K to 300 K with water which is available at 290 K. In the heat exchanger, tubes of 25 mm external and 22 mm internal diameter are employed and the water passes through the tubes. If the film coefficients for the water and benzene are 0.85 and 1.70 kW/m^2 K respectively and the scale resistance can be neglected, what total length of tube will

be required if the minimum quantity of water is to be used and its temperature is
not to be allowed to rise above 320 K?

Solution

Heat load:

On the benzene side, $Q = 1.25 \times 1.9(350 - 300) = 118.75$ kW.

In order to use the minimum amount of water, it must leave the unit at the
maximum temperature, i.e. 320 K. Thus for m kg/s water:

$$118.75 = m \times 4.18(320 - 290) \quad \text{or} \quad m = 0.947 \text{ kg/s}$$

Temperature driving force

$$\theta_1 = (350 - 320) = 30 \text{ K}, \quad \theta_2 = (300 - 290) = 10 \text{ K}$$

and in equation 7.10, $\theta_m = (30 - 10)/\ln(30/10) = 18.2$ K.

It will be assumed that the correction factor is unity.

Overall coefficient

Inside: $h_i = 0.85$ kW/m^2 K
or based on the tube o.d., $h_{io} = 0.85 \times 22/25 = 0.748$ kW/m^2 K.
Outside: $h_o = 1.70$ kW/m^2 K.
Wall: Taking $k_{steel} = 45$ W/m K

$$x/k = 0.003/45 = 0.00007 \text{ m}^2 \text{ K/W} \quad \text{or} \quad 0.07 \text{ m}^2 \text{ K/kW}$$

Thus neglecting any scale resistance,

$$1/U = (1/0.748) + (1/1.70) + 0.07 = 1.995$$

and $$U = 0.501 \text{ kW/m}^2 \text{ K}$$

Area

In equation 7.9: $A = Q/U\theta_m = 118.75/(0.0501 \times 18.2) = 13.02$ m^2.

Surface area of a 0.025 m o.d. tube $= (\pi \times 0.025 \times 1.0) = 0.0785$ m^2/m
and hence total length of tubing required $= (1302/0.0785)$
$$= 165.8 \text{ m}$$

Problem 7.38

Calculate the rate of loss of heat from a 6 m long horizontal steam pipe of 50 mm
internal diameter and 60 mm external diameter when carrying steam at 800 kN/m^2.
The temperature of the atmosphere and surroundings is 290 K.

What would be the cost of steam saved by coating the pipe with a 50 mm thickness of 85% magnesia lagging of thermal conductivity 0·07 W/m K, if steam costs £0·5 per 100 kg? The emissivity of the surface of the bare pipe and of the lagging may be taken as 0·85, and the coefficient h for the heat loss by natural convection can be calculated from the expression:

$$h = 1·65(\Delta T)^{0·25} \text{ (W/m}^2 \text{ K)}$$

where ΔT is the temperature difference in K.

Take the Stefan–Boltzmann constant as $5·67 \times 10^{-8}$ W/m² K⁴.

Solution

For the bare pipe

Steam is saturated at 800 kN/m² and 443 K.

Neglecting the inside resistance and that of the wall, it may be assumed that the surface temperature of the pipe is 443 K.

For *radiation* from the pipe, the surface area $= (\pi \times 0·060 \times 6·0) = 1·131$ m²
and in equation 7.84: $q_r = 5·67 \times 10^{-8} \times 0·85 \times 1·131(443^4 - 290^4)$
$$= 1714 \text{ W}$$

For *convection* from the pipe, the heat loss,

$$q_c = h_c A(T_s - T)$$
$$= 1·64(443 - 290)^{0·25} \times 1·131(443 - 290)$$
$$= 1·855(443 - 290)^{1·25} = 998 \text{ W}$$

and the total loss $= \underline{\underline{2712 \text{ W}}}$.

For the insulated pipe

The heat conducted through the lagging q_l must equal the heat lost from the surface $(q_r + q_c)$.

Mean diameter of the lagging $= [(0·060 + 2 \times 0·050) + 0·060]/2 = 0·110$ m
at which the area $= (\pi \times 0·110 \times 6·0) = 2·07$ m²
and in equation 7.11: $q_l = 0·07 \times 2·07(443 - T_s)/0·050 = (1280 - 2·90T_s)$ W
where T_s is the surface temperature.

The outside area is now $= \pi(0·060 + 2 \times 0·050) \times 6·0 = 3·016$ m²
and from equation 7.84: $q_r = 5·67 \times 10^{-8} \times 0·85 \times 3·016(T_s^4 - 290^4)$

$$= 1·456 \times 10^{-7}T_s^4 - 1030 \text{ W}$$

and
$$q_c = 1·64(T_s - 290)^{0·25} \times 3·016(T_s - 290)$$
$$= 4·946(T_s - 290)^{1·25} \text{ W}$$

Making a balance:

$$1280 - 2·90T_s = 1·456 \times 10^{-7}T_s^4 - 1030 + 4·946(T_s - 290)^{1·25}$$
or
$$4·946(T_s - 290)^{1·25} + 1·456 \times 10^{-7}T_s^4 + 2·90T_s = 2310$$

Solving by trial and error, $T_s = 305$ K
and hence the heat lost $= (1280 - 2.90 \times 305) = 396$ W.

The heat saved by lagging the pipe is thus $(2712 - 396) = 2317$ kW or 2.317 kW.

At 800 kN/m^2, the latent heat of steam is 2050 kJ/kg
and the reduction in the amount of steam condensed $= (2.317/2050) = 0.00113$ kg/s

or $\qquad (0.00113 \times 3600 \times 24 \times 365) = 35{,}643$ kg/year

$\therefore \qquad$ annual saving $= (35{,}643 \times 0.5/100) = \underline{£178/\text{year}}$

[*Note*: The concept of an arithmetic mean radius should only be used with thin walled tubes, which is not the case here, and the use of a logarithmic mean radius is of greater validity. If equation 7.8 is applied however, $T_s = 305.7$ K and the difference is negligible.]

Problem 7.39

A stirred reactor contains a batch of 700 kg reactants of specific heat 3.8 kJ/kg K initially at 290 K, which is heated by dry saturated steam at 170 kN/m^2 fed to a helical coil. During the heating period the steam supply rate is constant at 0.1 kg/s and condensate leaves at the temperature of the steam. If heat losses are neglected, calculate the true temperature of the reactants when a thermometer immersed in the material reads 360 K. The bulb of the thermometer is approximately cylindrical and is 100 mm long by 10 mm diameter with a water equivalent of 15 g, and the overall heat transfer coefficient to the thermometer is 300 W/m^2 K. What would a thermometer with a similar bulb of half the length and half the heat capacity indicate under these conditions?

Solution

The latent heat of dry saturated steam at 170 kN/m^2 and 388 K $= 2216$ kJ/kg.

Therefore heat added to the reactor $= (2216 \times 0.1) = 221.6$ kJ/s
which is equal to the increase in enthalpy, dH/dt.

The enthalpy of the contents, neglecting the heat capacity of the reactor and losses $\qquad = mC_p\, dT/dt$

$\qquad\qquad\qquad = (700 \times 3.8)\, dT/dt \quad$ or $\quad 2660\, dT/dt$ kJ/s

$\therefore \qquad\qquad 2660\, dT/dt = 221.6$

and the rate of temperature rise, $dT/dt = 0.083$ K/s.

At time t s, the temperature of the reactants is therefore

$$T = (290 + 0.083t)\ \text{K} \tag{i}$$

The increase in enthalpy of the thermometer is equal to the rate of heat transfer from the fluid, or

$$(mC_p)_t\, dT_t/dt = UA_t(T - T_t) \tag{ii}$$

where the subscript t refers to the thermometer.

$\therefore$ $(15/1000) \times 4 \cdot 18 (\mathrm{d}T_t/\mathrm{d}t) = 0 \cdot 300 (\pi \times 0 \cdot 010 \times 0 \cdot 100)(T - T_t)$

and $(\mathrm{d}T_t/\mathrm{d}t) = 0 \cdot 0150 (T - T_t) \ \mathrm{K/s}$

At time t s, the temperature of the thermometer is, therefore:

$$T_t = 290 + [0 \cdot 0150 (T - T_t)]t \ \mathrm{K} \tag{iii}$$

When $T_t = 360$ K, substituting from (i),

$$360 = 290 + \{0 \cdot 0150[(290 + 0 \cdot 083t) - 360]\}t$$

or $0 \cdot 00125 t^2 - 1 \cdot 05 - 70 = 0$

and $t = 902$ s

Therefore in (i), $T = (290 + 0 \cdot 083 \times 902)$

$$= \underline{\underline{364 \cdot 9 \ \mathrm{K}}}$$

With half the length, i.e. $0 \cdot 050$ m and half the heat capacity, i.e. $7 \cdot 5$ g water, in (ii):

$$(7 \cdot 5/1000) \times 4 \cdot 18 (\mathrm{d}T_t/\mathrm{d}t) = 0 \cdot 300 (\pi \times 0 \cdot 010 \times 0 \cdot 050)(T - T_t)$$

or $(\mathrm{d}T_t/\mathrm{d}t) = 0 \cdot 0150 (T - T_t)$

The same result as before and hence the new thermometer would also read $\underline{360 \ \mathrm{K}}$.

Problem 7.40

Derive an expression relating the pressure drop for the turbulent flow of a fluid in a pipe to the heat transfer coefficient at the walls on the basis of the simple Reynolds analogy. Indicate the assumptions which are made and the conditions under which you would expect it to apply closely. Air at 320 K and atmospheric pressure is flowing through a smooth pipe of 50 mm internal diameter and the pressure drop over a 4 m length is found to be 150 mm water gauge. By how much would you expect the air temperature to fall over the first metre if the wall temperature there is 290 K?

> Viscosity of air $= 0 \cdot 018$ mN s/m^2
>
> Specific heat $(C_p) = 1 \cdot 05$ kJ/kg K
>
> Kilogram molecular volume $= 22 \cdot 4$ m^3 at STP

Solution

If a mass of fluid, m kg, situated at a distance from a surface, is moving parallel to the surface with a velocity of u_s m/s, and it then moves to the surface, where the velocity is zero, it will give up its momentum (mu_s) kg m/s in time t s. If the temperature difference between the mass of fluid and the surface is θ_s, then the heat transferred to the surface is $(mC_p\theta_s)$ and over a surface of area, A,

$$(mC_p\theta_s)/t = -qA$$

where q is the heat transferred from the surface per unit area per unit time.

If the shear stress at the surface is R_o, the shearing force over area A is the rate of change of momentum or

$$(mu_s)/t = R_o A$$

$$\therefore \qquad C_p \theta_s / u_s = q / R_o$$

Writing $R_o = -R$, the shear stress acting on the walls and h as the heat transfer coefficient between the fluid and the surface:

$$-q/\theta = h = -R_o C_p / u_s = R C_p / u_s$$

or

$$h/C_p \rho u_s = R/\rho u^2$$

From equation 3.16, the pressure change due to friction is given by

$$\Delta P = 4(R/\rho u^2)(l/d)(\rho u^2) \tag{i}$$

and substituting from equation 10.38,

$$\Delta P = 4(h/C_p \rho u)(l/d)(\rho u^2)$$

$$= 4(hu/C_p)(l/d) \tag{ii}$$

The Reynolds analogy assumes no mixing with adjacent fluid and that turbulence persists right up to the surface. Further it is assumed that thermal and kinematic equilibria are reached when an element of fluid comes into contact with a solid surface. No allowance is made for variations in physical properties of the fluid with temperature.

A further discussion of the Reynolds analogy for heat transfer is presented in Chapter 10, page 348.

Density of air at 320 K = $(29/22 \cdot 4)(273/320) = 1 \cdot 105$ kg/m^3.

For pressure drop, $\Delta P = 150$ mm water = $(9 \cdot 8 \times 150) = 1470$ N/m^2

$$l = 4 \cdot 0 \text{ m}, \ d = 0 \cdot 050 \text{ m}$$

In equation 3.20:

$$\Delta P d^3 \rho / (4l\mu^2) = (1470 \times 0 \cdot 050^3 \times 1 \cdot 105)/[4 \times 4 \cdot 0 (0 \cdot 018 \times 10^{-3})^2]$$

$$= 3 \cdot 192 \times 10^7$$

From figure 3.8, for a smooth pipe, $Re = 1 \cdot 25 \times 10^5$
and from figure 3.7, $R/\rho u^2 = 0 \cdot 0021$

The heat transfer coefficient,

$$h = (R/\rho u^2)C_p \rho u = 0 \cdot 0021 \times 1 \cdot 05 \times 10^3 \rho u = 2 \cdot 205 \rho u \text{ W/m}^2 \text{ K}$$

Mass flow of air, $\quad m = \rho u (\pi/4)0 \cdot 050^2 = 0 \cdot 00196 \rho u$ kg/s

Area for heat transfer, $A = (\pi \times 0 \cdot 050 \times 1 \cdot 0) = 0 \cdot 157$ m^2.

Now $\qquad\qquad\qquad m C_p (T_1 - T_2) = hA(T_m - T_w)$

where T_1 and T_2 are the inlet and outlet temperatures and T_m the mean value taken as arithmetic over the small length of 1 m.

$\therefore \quad 0 \cdot 00196 \rho u \times 1 \cdot 050 \times 10^3 (320 - T_2) = 2 \cdot 205 \rho u \times 0 \cdot 157(0 \cdot 5(320 + T_2) - 290)$

and $\qquad\qquad\qquad\qquad T_2 = 316$ K

The drop in temperature over the first metre is therefore $\underline{\underline{4 \text{ K}}}$.

Problem 7.41

The radiation received by the earth's surface on a clear day with the sun overhead is $1\,kW/m^2$ and an additional $0\cdot3\,kW/m^2$ is absorbed by the earth's atmosphere. Calculate approximately the temperature of the sun, assuming its radius to be 700,000 km and the distance between the sun and the earth to be 150,000,000 km. The sun may be assumed to behave as a black body.

Solution

The total radiation received $= 1\cdot3\,kW/m^2$ of the earth's surface. The equivalent surface area of the sun is obtained by comparing the area of a sphere at the radius of the sun, 7×10^5 km and the area of a sphere of radius (radius of sun + distance between sun and earth) or

$$A_1/A_2 = 4\pi(7 \times 10^5)^2/4\pi(150 \times 10^6 + 7 \times 10^5)^2$$
$$= 2\cdot16 \times 10^{-5}$$

Therefore radiation at the sun's surface $= 1\cdot3 \times 10^3/2\cdot16 \times 10^{-5} = 6\cdot03 \times 10^7\,W/m^2$. For a black body, the intensity of radiation is given by equation 7.82,

or $$6\cdot03 \times 10^7 = 5\cdot67 \times 10^{-8}T^4$$
$\therefore$ $$\underline{\underline{T = 5710\,K}}$$

Problem 7.42

A thermometer is immersed in a liquid which is heated at the rate of $0\cdot05\,K/s$. If the thermometer and the liquid are both initially at 290 K, what rate of passage of liquid over the bulb of the thermometer is required if the error in the thermometer reading after 600 s is to be no more than 1 K? Take the water equivalent of the thermometer as 30 g and the heat transfer coefficient to the bulb to be given by $U = 735\,u^{0\cdot8}\,W/m^2\,K$, and the area of the bulb as $0\cdot01\,m^2$ where u is the velocity in m/s.

Solution

Let T and T' be the liquid and thermometer temperatures respectively after time t s.

$$dT/dt = 0\cdot05\,K/s \quad \text{and hence} \quad T = 290 + 0\cdot05t$$

When $t = 600$ s, $(T - T') = 1$.
$\therefore$ $$T = 290 + (600 \times 0\cdot05) = 320\,K \quad \text{and} \quad T' = 319\,K$$
Making a balance,

$$m_t Cp_t\,dT'/dt = UA(T - T')$$
$\therefore$ $$(30/1000)4\cdot18\,dT'/dt = UA(290 + 0\cdot05t - T')$$
$\therefore$ $$dT'/dt + 7\cdot98\,UAT' = 2312\,UA(1 + 0\cdot000173t)$$

$$\therefore \qquad e^{7\cdot98UAt}T' = 2312\,UA\left\{(1+0\cdot000173t)\frac{e^{7\cdot98UAt}}{7\cdot98\,UA} - \int 0\cdot000173\,\frac{e^{7\cdot98UAt}}{7\cdot98\,UA}\,dt\right\}$$

$$= 290(1+0\cdot000173t)\,e^{7\cdot98UAt} - 0\cdot050\,\frac{e^{7\cdot98UAt}}{7\cdot98\,UA} + k$$

When $t = 0$, $T' = 290$ K. $\therefore k = 0\cdot00627/UA$.

$$\therefore \qquad T' = 290(1+0\cdot000173t) - (0\cdot00627/UA)(1 - e^{-7\cdot98UAt})$$

When $t = 600$ s, $T' = 319$ K.

$$\therefore \qquad 319 = 320 - (0\cdot00627/UA)(1 - e^{\,4789UA})$$

$$\therefore \qquad -4789\,UA = \ln(1 - 159\cdot5\,UA)$$

and $\qquad\qquad UA = -0\cdot000209\ln(1 - 159\cdot5\,UA)$

Solving by trial and error, $UA = 0\cdot00627$ kW/K.

$$A = 0\cdot01\ \text{m}^2 \quad \text{and hence} \quad U = 0\cdot627\ \text{kW/m}^2\,\text{K or } 627\ \text{W/m}^2\,\text{K}$$

$$\therefore \qquad\qquad 627 = 735u^{0\cdot8}$$

and $\qquad\qquad \underline{\underline{u = 0\cdot82\ \text{m/s}}}$

Problem 7.43

In a shell-and-tube type of heat exchanger with horizontal tubes 25 mm external diameter and 22 mm internal diameter, benzene is condensed on the outside by means of water flowing through the tubes at the rate of $0\cdot03$ m³/s. If the water enters at 290 K and leaves at 300 K and the heat transfer coefficient on the water side is 850 W/m² K, what total length of tubing will be required?

Solution

Heat load: mass flow of water $= (0\cdot03 \times 1000) = 30$ kg/s
and hence the heat load $= 30 \times 4\cdot18(300 - 290) = 1254$ kW
At atmospheric pressure, benzene condenses at 353 K and hence

$$\theta_1 = (353 - 290) = 63\ \text{K}, \quad \theta_2 = (353 - 300) = 53\ \text{K}$$

and from equation 7.10,

$$\theta_m = (63 - 53)/\ln(63/53) = 57\cdot9\ \text{K}$$

No correction factor is required.

For condensing benzene, h_o will be taken as 1750 W/m² K. From Table 7.13, $h_i = 850$ W/m² K or, based on the outside diameter, $h_{io} = (850 \times 22/25) = 748$ W/m² K.

Neglecting scale and wall resistances,

$$1/U = 1/1750 + 1/748 = 0\cdot00191\ \text{m}^2\,\text{K/W}$$

and $\qquad\qquad U = 524\ \text{W/m}^2\,\text{K} \quad \text{or} \quad 0\cdot524\ \text{kW/m}^2\,\text{K}$

Therefore from equation 7.9: $A = 1254/(0\cdot524 \times 57\cdot9) = 41\cdot3$ m².
Outside area of $0\cdot025$ m tubing $= (\pi \times 0\cdot025 \times 1\cdot0) = 0\cdot0785$ m²/m
and total length of tubing required $= (41\cdot3/0\cdot0785) = \underline{526\ \text{m}}$.

Problem 7.44

In a contact sulphuric acid plant, the gases leaving the first convertor are to be cooled from 845 to 675 K by means of the air required for the combustion of the sulphur. The air enters the heat exchanger at 495 K. If the flow of each of the streams is 2 m³/s at NTP, suggest a suitable design for a shell-and-tube type of heat exchanger employing tubes of 25 mm internal diameter.

(a) Assume parallel co-current flow of the gas streams.

(b) Assume parallel countercurrent flow.

(c) Assume that the heat exchanger is fitted with baffles giving cross-flow outside the tubes.

Solution

Heat load

At a mean air temperature of 288 K (NTP), density $= (29/22\cdot4)(273/288) = 1\cdot227$ kg/m³.

Therefore mass flow of air $= (2\cdot0 \times 1\cdot227) = 2\cdot455$ kg/s.

If, as a first approximation, the thermal capacities of the two streams can be assumed equal for equal flowrates, then the outlet air temperature $= 495 + (845 - 675) = 665$ K and for a mean specific heat of $1\cdot0$ kJ/kg K, the heat load is given by:

$$Q = 2\cdot455 \times 1\cdot0(665 - 495) = 417\cdot4\ \text{kW}$$

For gas to gas heat transfer, an overall coefficient of $1/(1/60 + 1/60) = 30$ W/m² K will be taken from Table 7.13.

(a) *Co-current flow*

$$\theta_1 = (845 - 495) = 350\ \text{K}, \quad \theta_2 = (675 - 665) = 10\ \text{K}$$

and in equation 7.10: $\theta_m = (350 - 10)/\ln(350/10) = 95\cdot6$ K.

Therefore in equation 7.9, $A = (417\cdot4 \times 10^3)/(30 \times 95\cdot6) = 145\cdot5$ m².

For 25 mm i.d. tubes an o.d. of 32 mm will be assumed for which the outside area $= (\pi \times 0\cdot032 \times 1\cdot0) = 0\cdot1005$ m²/m and total length of tubing $= (145\cdot5/0\cdot1005) = 1447$ m.

At a mean air temperature of 580 K, $\rho = (29/22\cdot4)(273/580) = 0\cdot609$ kg/m³.

Therefore volume flow of air $= (2\cdot455/0\cdot609) = 4\cdot03$ m³/s.

For a reasonable gas velocity of say 15 m/s, area for flow $= (4\cdot03/15) = 0\cdot268$ m².

Cross-sectional area of one tube $= (\pi/4)0\cdot025^2 = 0\cdot00050$ m².

Hence number of tubes/pass $= (0\cdot268/0\cdot00050) = 545$ each of length $= (1447/545) = 2\cdot65$ m.

In practice, the standard length of $2\cdot44$ m would be adopted with $(1447/2\cdot44) = 594$ tubes in a single pass.

(b) *Counter flow*

$$\theta_1 = (845 - 665) = 180 \text{ K}, \quad \theta_2 = (675 - 495) = 180 \text{ K}, \quad \text{and} \quad \theta_m = 180 \text{ K}$$

$\therefore$ in equation 7.9: $A = (417 \cdot 4 \times 10^3)/(30 \times 180) = 77 \cdot 3 \text{ m}^2$
and total length of tubing $= (77 \cdot 3/0 \cdot 1005) = 769 \text{ m}$.

With a velocity of 15 m/s, each tube would be $(769/545) = 1 \cdot 41 \text{ m}$ long.

Possibly a better arrangement would be the use of $(769/2 \cdot 44) = 315$ tubes, $2 \cdot 44$ m long, though this would give a higher velocity and hence an increased air side pressure drop. With such an arrangement, 315×32 mm o.d. tubes could be accommodated in a 838 mm i.d. shell on 40 mm triangular pitch.

(c) *Cross flow*

As in (b): $\theta_m = 180 \text{ K}$

$$X = (t_2 - t_1)/(T_1 - t_1) = (665 - 495)/(845 - 495) = 0 \cdot 486$$
and $\qquad Y = (T_1 - T_2)/(t_2 - t_1) = (845 - 675)/(665 - 495) = 1 \cdot 0$

Thus, assuming one shell pass, two tube side passes, from Figure 7.51, $F = 0 \cdot 82$ and $\theta_m F = (0 \cdot 82 \times 180) = 147 \cdot 6 \text{ K}$.

Thus, in equation 7.140: $A = (417 \cdot 4 \times 10^3)/(30 \times 147 \cdot 6) = 94 \cdot 3 \text{ m}^2$ and total length of tubing $= (94 \cdot 3/0 \cdot 1005) = 938 \text{ m}$.

Using standard tubes $2 \cdot 44$ m long, number of tubes $= (938/2 \cdot 44) = 384$ or $(384/2) = 192$ tubes/pass.

The cross-sectional area for flow would then be $(192 \times 0 \cdot 00050) = 9 \cdot 61 \times 10^{-2} \text{ m}^2$ and the air velocity $= (4 \cdot 03/9 \cdot 61 \times 10^{-2}) = 41 \cdot 9 \text{ m/s}$.

This is not excessive providing the minimum acceptable pressure drop is not exceeded.

The nearest standard size is 390×32 mm o.d. tubes, $2 \cdot 44$ m in a 940 mm i.d. shell arranged in two passes on 40 mm triangular pitch.

Problem 7.45

A large block of material of thermal diffusivity $D_H = 0 \cdot 0042 \text{ cm}^2/\text{s}$ is initially at a uniform temperature of 290 K and one face is raised suddenly to 875 K and maintained at that temperature. Calculate the time taken for the material at a depth of $0 \cdot 45$ m to reach a temperature of 475 K on the assumption of unidirectional heat transfer and that the material can be considered to be infinite in extent in the direction of transfer.

Solution

This problem is identical to Problem 7.2 except for slight variations in temperature, and reference should be made to that solution.

Problem 7.46

A 50% glycerol–water mixture is flowing at a Reynolds number of 1500 through a 25 mm diameter pipe. Plot the mean value of the heat transfer coefficient as a function of pipe length, assuming that

$$Nu = 1 \cdot 62 (Re \, Pr \, d/L)^{0 \cdot 33}$$

Indicate the conditions under which this is consistent with the predicted value $Nu = 4 \cdot 1$ for fully developed flow.

Solution

For 50% glycerol-water at, say, 290 K, $\mu = 0 \cdot 007$ N s/m², $k = 0 \cdot 415$ W/m K, and $C_p = 3135$ J/kg K.

$$\therefore \qquad h \times 0 \cdot 025/0 \cdot 415 = 1 \cdot 62[(1500 \times 3135 \times 0 \cdot 007/0 \cdot 415)(0 \cdot 025/L)]^{0 \cdot 33}$$

$$\therefore \qquad h = 26 \cdot 89(1983/L)^{0 \cdot 33}$$

$$= 330/L^{0 \cdot 33} \text{ W/m}^2 \text{ K}$$

h is plotted as a function of L over the range $L = 0$–10 m in Figure 7d. When $Nu = 4 \cdot 1$, $h = 4 \cdot 1 k/d = (4 \cdot 1 \times 0 \cdot 415/0 \cdot 025) = 68 \cdot 1$ W/m² K.

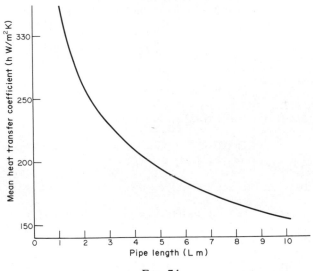

FIG. 7d

Taking this as a point value,

$$L = (330/68 \cdot 1)^3 = \underline{113 \cdot 8 \text{ m}}$$

which would imply that the flow is fully developed at this point. For further discussion on this point reference should be made to turbulent flow of gases in Section 7.4.3.

Problem 7.47

A liquid is boiled at a temperature of 360 K using steam fed at a temperature of 380 K to a coil heater. Initially the heat transfer surfaces are clean and an evaporation rate of 0·08 kg/s is obtained from each square metre of heating surface. After a period, a layer of scale of resistance 0·0003 m^2 K/W, is deposited by the boiling liquid on the heat transfer surface. On the assumption that the coefficient on the steam side remains unaltered and that the coefficient for the boiling liquid is proportional to its temperature difference raised to the power of 2·5, calculate the new rate of boiling.

Solution

When the surface is clean, taking the wall and the inside resistances as negligible, the surface temperature will be 380 K.

Thus
$$Q = h_o A (T_s - T)$$

where $Q = mL$ (m kg/s is the rate of evaporation of fluid of latent heat L J/kg), $A = 1$ m^2, and T_s and T are the surface and fluid temperature respectively.

$\therefore$
$$0 \cdot 08 L = h_o \times 1 \cdot 0 (380 - 360)$$

or
$$h_o = 0 \cdot 004 L$$

Now
$$h_o \propto (T_s - T)^{2 \cdot 5}$$

or
$$h_o = k'(380 - 360)^{2 \cdot 5} = 1 \cdot 79 \times 10^3 k'$$

and
$$k' = 0 \cdot 004 L / 1 \cdot 79 \times 10^3 = 2 \cdot 236 \times 10^{-6} L$$

When the scale has formed, the total resistance
$$= 0 \cdot 0003 + 1/(2 \cdot 236 \times 10^{-6} L (T_s - 360)^{2 \cdot 5})$$
$$= 0 \cdot 0003 + 4 \cdot 472 \times 10^5 L^{-1} (T_s - 360)^{-2 \cdot 5}$$

For conduction through the scale:
$$mL = (380 - T_s)/0 \cdot 0003 = 3 \cdot 33 \times 10^3 (380 - T_s) \tag{i}$$

For transfer through the outside film:
$$mL = (t - 360)/[4 \cdot 472 \times 10^5 L^{-1} (T_s - 360)^{-2 \cdot 5}] = 2 \cdot 236 \times 10^{-6} L (T_s - 360)^{3 \cdot 5} \tag{ii}$$

and for overall transfer:
$$mL = (380 - 360)/[0 \cdot 0003 + 4 \cdot 472 \times 10^5 L^{-1} (T_s - 360)^{-2 \cdot 5}] \tag{iii}$$

Inspection of these equations shows that the rate of evaporation m is a function not only of the surface temperature T_s but also of the latent heat of the fluid L. Using equations (i) and (ii) and selecting values of T in the range 360 to 380 K, the following results are obtained:

Surface temperature T_s (K)	Mass rate of evaporation m (kg/s)	Latent heat of vaporisation L (kJ/kg)
362	0·000025	2400000
364	0·00029	186000
366	0·0012	39600
368	0·0033	12200
370	0·0071	4710
372	0·013	1990
374	0·023	869
376	0·036	364
378	0·055	121
380	0·080	0

At a boiling point of 360 K it is likely that the liquid is organic with a latent heat of, say, 900 kJ/kg. This would indicate a surface temperature of 374 K and an evaporation rate of 0·023 kg/s. A precise result requires more specific data on the latent heat.

Problem 7.48

A batch of reactants of specific heat 3·8 kJ/kg K and weighing 1000 kg is heated by means of a submerged steam coil of area 1 m², fed with steam at 390 K. If the overall heat transfer coefficient is 600 W/m² K, calculate the time taken to heat the material from 290 to 360 K if heat losses to the surroundings are neglected.

If the external area of the vessel is 10 m² and the heat transfer coefficient to the surroundings at 290 K is 8·5 W/m² K, what will be the time taken to heat the reactants over the same temperature range and what is the maximum temperature to which the reactants can be raised?

What methods would you suggest for improving the rate of heat transfer?

Solution

Use is made of equation 7.138:

$$\ln[(T_s - T_1)/(T_s - T_2)] = UAt/mC_p$$

∴
$$\ln[(390 - 290)/(390 - 360)] = 600 \times 1 \cdot 0t/(1000 \times 3 \cdot 8 \times 10^3)$$

or
$$\ln 3 \cdot 33 = 0 \cdot 000158t$$

and
$$\underline{t = 7620 \text{ s}}$$

The heat lost from the vessel, $Q_L = hA_v(T - T_a)$, where T_a is ambient temperature

∴
$$Q_L = 8 \cdot 5 \times 10 \cdot 0(T - 290)$$

$$= 85 \cdot 0T - 24650 \text{ W}$$

The heat from the steam = heat to the reactants + heat losses

$$UA(T_s - T) = mC_p \, dT/dt + 85 \cdot 0T - 24650$$

$$600 \times 1 \cdot 0(390 - T) = 1000 \times 3 \cdot 8 \times 10^3 \, dT/dt + 85 \cdot 0T - 24650$$

$$\int_0^t dt = 5548 \int_{T_1}^{T_2} dT/(377 \cdot 6 - T)$$

$$\therefore \qquad t = 5548 \ln[(377 \cdot 6 - T_1)/(377 \cdot 6 - T_2)]$$

$$= 5548 \ln[(377 \cdot 6 - 290)/(377 \cdot 6 - 360)]$$

$$= \underline{\underline{8904 \text{ s}}}$$

The maximum temperature of the reactants is attained when the heat transferred from the steam is equal to the heat losses, or:

$$UA(T_s - T) = hA_v(T - T_a)$$

$$600 + 1 \cdot 0(390 - T) = 8 \cdot 5 \times 10 \cdot 0(T - 290)$$

or $\qquad\qquad\qquad\qquad \underline{\underline{T = 378 \text{ K}}}$

The heating-up time could be reduced by improving the rate of heat transfer to the fluid, by agitation of the fluid for example and by reducing heat losses from the vessel by insulation. In the case of a large vessel there is a limit to the degree of agitation and circulation of the fluid through an external heat exchanger is an attractive possibility.

Problem 7.49

What do you understand by the terms "black body" and "grey body" when applied to radiant heat transfer?

Two large parallel plates with grey surfaces are situated 75 mm apart; one has an emissivity of $0 \cdot 8$ and is at a temperature of 350 K and the other has an emissivity of $0 \cdot 4$ and is at a temperature of 300 K. Calculate the net rate of heat exchange by radiation per square metre taking the Stefan–Boltzmann constant as $5 \cdot 67 \times 10^{-8} \text{ W/m}^2 \text{ K}^4$. Any formula (other than Stefan's law) which you use must be proved.

Solution

The terms "black body" and "grey body" are discussed in Sections 7.5.1 and 7.5.2.

For two large parallel plates with grey surfaces, the heat transfer by radiation between them is given by equation 7.89:

$$q = [e_1 e_2 \sigma/(e_1 + e_2 - e_1 e_2)](T_1^4 - T_2^4) \text{ W/m}^2$$

This equation is derived by the multiple reflection method in Section 7.5.4 and by the net radiation method also in Section 7.5.4.

In this case:

$$q = [0{\cdot}8 \times 0{\cdot}4 \times 5{\cdot}67 \times 10^{-8}/(0{\cdot}8 + 0{\cdot}4 - 0{\cdot}8 \times 0{\cdot}4)](350^4 - 300^4)$$
$$= 0{\cdot}367 \times 5{\cdot}67 \times 10^{-8} \times 6{\cdot}906 \times 10^9$$
$$= \underline{\underline{143{\cdot}7 \text{ W/m}^2}}$$

Problem 7.50

A longitudinal fin on the outside of a circular pipe is 75 mm deep and 3 mm thick. If the pipe surface is at 400 K, calculate the heat dissipated per metre length from the fin to the atmosphere at 290 K, if the coefficient of heat transfer from its surface may be assumed constant at 5 W/m^2 K. The thermal conductivity of the material of the fin is 50 W/m K and the heat loss from the extreme edge of the fin may be neglected. It should be assumed that the temperature is uniformly 400 K at the base of the fin.

Solution

The heat lost from the fin is given by equation 7.150:

$$Q_f = \sqrt{(hbkA)}\theta_1 \tanh mL$$

where h is the coefficient of heat transfer to the surroundings $= 5$ W/m^2 K, b is the fin perimeter $= (2 \times 0{\cdot}075 + 0{\cdot}003) = 0{\cdot}153$ m, k is the thermal conductivity of the fin $= 50$ W/m K, A is the cross-sectional area of the fin $= (0{\cdot}003 \times 1{\cdot}0) = 0{\cdot}003$ m^2, θ_1 is the temperature difference at the root $= (T_1 - T_G) = (400 - 290) = 110$ K, $m = \sqrt{(hb/kA)} = \sqrt{((5 \times 0{\cdot}153)/(50 \times 0{\cdot}003))} = 2{\cdot}258$, and L is the length of the fin $= 0{\cdot}075$ m.

$$\therefore \qquad Q_f = \sqrt{(5 \times 0{\cdot}153 \times 50 \times 0{\cdot}003)} \times 110 \tanh (2{\cdot}258 - 0{\cdot}075)$$
$$= (0{\cdot}339 \times 110 \tanh 0{\cdot}169) = \underline{\underline{6{\cdot}23 \text{ W/m}}}$$

Problem 7.51

Liquid oxygen is distributed by road in large spherical insulated vessels, 2 m internal diameter, well lagged on the outside. What thickness of magnesia lagging, of thermal conductivity 0·07 W/m K, must be used so that not more than 1% of the liquid oxygen evaporates during a journey of 10 ks (2·78 h) if the vessel is initially 80% full?

Latent heat of vaporization of oxygen	$= 215$ kJ/kg
Boiling point of oxygen	$= 90$ K
Density of liquid oxygen	$= 1140$ kg/m^3
Atmospheric temperature	$= 288$ K
Heat transfer coefficient from outside lagging to atmosphere.	$= 4{\cdot}5$ W/m^2 K

Solution

For conduction through the lagging, equation 7.21 may be used:

$$Q = 4\pi k(T_1 - T_2)/(1/r_1 - 1/r_2)$$

where T_1 will be taken as the temperature of boiling oxygen = 90 K and r_1 is the tank radius = 1·0 m.

In this way, the resistance to heat transfer in the inside film and the walls is neglected. r_2 is the outer radius of the lagging.

$\therefore$
$$Q = 4\pi \times 0.07(90 - T_2)/(1/1.0 - 1/r_2) \quad \text{W} \tag{i}$$

For heat transfer from the outside of the lagging to the surroundings:

$$Q = hA(T_2 - T_a)$$

where $h = 4.5 \text{ W/m}^2 \text{ K}$, $A = 4\pi r_2^2$, and T_a, ambient temperature, = 288 K.

$\therefore$
$$Q = 4.5 \times 4\pi r_2^2(T_2 - 288)$$
$$= 18\pi r_2^2(T_2 - 288) \quad \text{W} \tag{ii}$$

The volume of the tank $= 4\pi r_1^3/3 = 4\pi \times 1.0^3/3 = 4.189 \text{ m}^3$.

$\therefore$ volume of oxygen $= (4.189 \times 80/100) = 3.351 \text{ m}^3$

and mass of oxygen $= (3.351 \times 1140) = 3820 \text{ kg}$

$\therefore$ mass of oxygen which evaporates $= (3820 \times 1/100) = 38.2 \text{ kg}$

or $38.2/(10 \times 10^3) = 0.00382 \text{ kg/s}$

$\therefore$ heat flow into vessel, $Q = (215 \times 10^3 \times 0.00382) = -821 \text{ W}$

$\therefore$ In (ii) $-821 = 18\pi r_2^2(T_2 - 288)$

$\therefore$
$$T_2 = 288 - (14.52/r_2^2)$$

Substituting in (i):

$$-821 = 4\pi \times 0.07[90 - 288 + (14.52/r_2^2)]/(1 - 1/r_2)$$

or $$r_2^2 - 1.27r_2 + 0.0198 = 0$$

and $$r_2 = 1.25 \text{ m}$$

Thus the thickness of lagging $= (r_2 - r_1) = \underline{\underline{0.25 \text{ m}}}$.

Problem 7.52

Benzene is to be condensed at the rate of 1·25 kg/s in a vertical shell and tube type of heat exchanger fitted with tubes of 25 mm outside diameter and 2·5 m long. The vapour condenses on the outside of the tubes and the cooling water enters at 295 K and passes through the tubes at 1·05 m/s. Calculate the number of tubes required if the heat exchanger is arranged for a single pass of the cooling water. The tube wall thickness is 1·6 mm.

Solution

Preliminary calculation

At $101 \cdot 3 \text{ kN/m}^2$, benzene condenses at 353 K at which the latent heat = 394 kJ/kg

∴ heat load, $Q = (1 \cdot 25 \times 394) = 492 \text{ kW}$

The maximum water outlet temperature to minimise scaling is 320 K and a value of 300 K will be selected. Thus the water flow is given by

$$492 = m \times 4 \cdot 18(300 - 295)$$

or $m = 23 \cdot 5 \text{ kg/s}$ or $(23 \cdot 5/1000) = 0 \cdot 0235 \text{ m}^3/\text{s}$

∴ area required for a velocity of $1 \cdot 05 \text{ m/s} = (0 \cdot 0235/1 \cdot 05) = 0 \cdot 0224 \text{ m}^2$

The cross-sectional area of a tube of $(25 - 2 \times 1 \cdot 6) = 21 \cdot 8 \text{ mm i.d.}$ is $(\pi/4) \times 0 \cdot 0218^2 = 0 \cdot 000373 \text{ m}^2$
and hence number of tubes required = $(0 \cdot 0224/0 \cdot 000373) = 60$ tubes.

The outside area = $(\pi \times 0 \cdot 025 + 2 \cdot 5 \times 60) = 11 \cdot 78 \text{ m}^2$

$$\theta_1 = (353 - 295) = 58 \text{ K}, \quad \theta_2 = (353 - 300) = 53 \text{ K}$$

and in equation 7.10: $\theta_m = (58 - 53)/\ln(58/53) = 55 \cdot 5 \text{ K}$

∴ $U = 492/(55 \cdot 5 \times 11 \cdot 78) = 0 \cdot 753 \text{ kW/m}^2 \text{ K}$

This is quite reasonable as it falls in the middle of the range for condensing organics as given in Table 7.12. It remains to check whether the required overall coefficient will be attained with this geometry.

Overall coefficient

Inside:

The simplified equation for water in tubes, equation 7.59 may be used:

$$h_i = 1063(1 + 0 \cdot 00293\,T)u^{0 \cdot 8}/d_i^{0 \cdot 2} \text{ W/m}^2 \text{ K}$$

where $T = 0 \cdot 5(300 + 295) = 297 \cdot 5 \text{ K}$
$u = 1 \cdot 05 \text{ m/s}$ and $d_i = 0 \cdot 0218 \text{ m}$

∴ $h_i = 1063(1 + 0 \cdot 00293 \times 297 \cdot 5)1 \cdot 05^{0 \cdot 8}/0 \cdot 0218^{0 \cdot 2} = 4447 \text{ W/m}^2 \text{ K}$

or $4 \cdot 45 \text{ kW/m}^2 \text{ K}$

Based on the outside diameter:

$$h_{io} = (4 \cdot 45 \times 0 \cdot 0218/0 \cdot 025) = 3 \cdot 88 \text{ kW/m}^2 \text{ K}$$

Wall:

For steel, $K = 45 \text{ W/m K}$, $x = 0 \cdot 0016 \text{ m}$, and hence

$$x/k = (0 \cdot 0016/45) = 0 \cdot 000036 \text{ m}^2 \text{ K/W}$$

or $0 \cdot 036 \text{ m}^2 \text{ K/kW}$

Outside:

For condensation on vertical tubes, equation 7.105 may be used:

$$h_o(\mu^2/k^3\rho^2g)^{0.33} = 1.47(4\,M/\mu)^{-0.33}$$

The wall temperature is approximately $0.5(353 + 297.5) = 325$ K, and the benzene film temperature will be taken as $0.5(353 + 325) = 339$ K.

At 339 K, $k = 0.15$ W/m K, $\rho = 880$ kg/m³, and $\mu = 0.35 \times 10^{-3}$ N s/m².

With 60 tubes, the mass flow of benzene per tube, $G' = (1.25/60) = 0.0208$ kg/s.

For vertical tubes, $M = G'/\pi d_o = 0.0208/(\pi \times 0.025) = 0.265$ kg/m s.

$$\therefore \quad h_o[(0.35 \times 10^{-3})^2/0.15^3 \times 880^2 \times 9.81]^{0.33} = 1.47[4 \times 0.0208/(0.35 \times 10^{-3})]^{-0.33}$$

$$\therefore \qquad\qquad 1.699 \times 10^{-4}h_o = 1.47 \times 1.62 \times 10^{-1}$$

and
$$h_o = 1399 \text{ W/m}^2\text{ K}$$

or
$$1.40 \text{ kW/m}^2\text{ K}$$

Overall:

Neglecting scale resistances:

$$1/U = 1/h_{io} + x/k + 1/h_o = 0.258 + 0.036 + 0.714 = 1.008 \text{ m}^2\text{ K/kW}$$

and
$$U = 0.992 \text{ kW/m}^2\text{ K}$$

This is in excess of the value required and would allow for a reasonable scale resistance. If this were negligible, the water throughput could be reduced.

On the basis of these calculations, <u>60 tubes are required</u>.

Problem 7.53

One end of a metal bar 25 mm in diameter and 0.3 m long is maintained at 375 K, and heat is dissipated from the whole length of the bar to surroundings at 295 K. If the coefficient of heat transfer from the surface is 10 W/m² K, what is the rate of loss of heat? Take the thermal conductivity of the metal as 85 W/m K.

Solution

Use is made of equation 7.150:

$$Q_f = \sqrt{(hbkA)}\,\theta_1\tanh mL$$

where the coefficient of heat transfer from the surface, $h = 10$ W/m² K; the perimeter, $b = (\pi \times 0.025) = 0.0785$ m; the cross-sectional area, $A = (\pi/4) \times 0.025^2 = 0.000491$ m²; the thermal conductivity of the metal, $k = 85$ W/m K; the temperature difference at the root, $\theta_1 = (375 - 295) = 80$ K; the value of $m = \sqrt{(hb/kA)} = \sqrt{[(10 \times 0.0785)/(85 \times 0.000491)]} = 4.337$, and the length of the rod, $L = 0.3$ m.

$$\therefore \qquad Q_f = \sqrt{(10 \times 0.0785 \times 85 \times 0.000491)} \times 80 \tanh(4.337 \times 0.3)$$
$$= \sqrt{(0.0328)} \times 80 \tanh 1.3011$$

$$= 14 \cdot 49(e^{1 \cdot 301} - e^{-1 \cdot 301})/(e^{1 \cdot 301} + e^{-1 \cdot 301})$$
$$= 14 \cdot 49(3 \cdot 673 - 0 \cdot 272)/(3 \cdot 673 + 0 \cdot 272)$$
$$= \underline{\underline{12 \cdot 5 \text{ W}}}$$

Problem 7.54

A shell-and-tube heat exchanger consists of 120 tubes of internal diameter 22 mm and length 2·5 m. It is operated as a single-pass condenser with benzene condensing at a temperature of 350 K on the outside of the tubes and water of inlet temperature 290 K passing through the tubes. Initially there is no scale on the walls, and a rate of condensation of 4 kg/s is obtained with a water velocity of 0·7 m/s through the tubes. After prolonged operation, a scale of resistance 0·0002 m^2 K/W is formed on the inner surface of the tubes. To what value must the water velocity be increased in order to maintain the same rate of condensation on the assumption that the transfer coefficient on the water side is proportional to the velocity raised to the 0·8 power, and that the coefficient for the condensing vapour is 2·25 kW/m^2 K, based on the inside area? The latent heat of vaporization of benzene is 400 kJ/kg.

Solution

Area for heat transfer, based on tube i.d. $= (\pi \times 0 \cdot 022 \times 1 \cdot 0) = 0 \cdot 0691$ m^2/m or $(120 \times 2 \cdot 5 \times 0 \cdot 0691) = 20 \cdot 74$ m^2.

With no scale

Heat load, $Q = (4 \times 400) = 1600$ W.

Cross-sectional area of one tube $= (\pi/4) \times 0 \cdot 022^2 = 0 \cdot 00038$ m^2 and hence area for flow per pass $= (120 \times 0 \cdot 00038) = 0 \cdot 0456$ m^2.

∴   volume of flow of water $= (0 \cdot 0456 \times 0 \cdot 7) = 0 \cdot 0319$ m^3/s

and   mass flow of water $= (0 \cdot 0319 \times 1000) = 31 \cdot 93$ kg/s

The water outlet temperature is given by:

$$1600 = 31 \cdot 93 \times 4 \cdot 18(t - 290)$$

or $$t = 302 \text{ K}$$

$$\theta_1 = (350 - 290) = 60 \text{ K}, \quad \theta_2 = (350 - 302) = 48 \text{ K}$$

and in equation 7.10: $\theta_m = (60 - 48)/\ln(60/48) = 53 \cdot 8$ K.

In equation 7.9: $U = Q/A\theta_m$

$$= 1600/(20 \cdot 74 \times 53 \cdot 8) = 1 \cdot 435 \text{ kW/}m^2 \text{ K}$$

Neglecting the wall resistance,

$$1/U = 1/h_i + 1/h_{oi}$$
$$1/1\cdot435 = 1/h_i + 1/2\cdot25$$
$$\therefore \qquad h_i = 3\cdot958 \text{ kW/m}^2 \text{ K}$$

h_i is proportional to $u^{0\cdot8}$ or $3\cdot958 = K(0\cdot7)^{0\cdot8}$
$$\therefore \qquad K = 5\cdot265$$

With scale

$$h_i = 5\cdot265u^{0\cdot8} \text{ kW/m}^2 \text{ K} \quad \text{(scale resistance} = 0\cdot20 \text{ m}^2 \text{ K/kW)}$$

and
$$1/U = 1/(5\cdot265u^{0\cdot8}) + 0\cdot20 + 1/2\cdot25$$

and
$$U = u^{0\cdot8}/(0\cdot190 + 0\cdot644u^{0\cdot8}) \text{ kW/m}^2 \text{ K}$$

$$Q = 1600 \text{ kW} \quad \text{as before}$$

The mass flow of water $= u \times 0\cdot0456 \times 1000 = 45\cdot6 \, u$ kg/s and the outlet water temperature is given by:

$$1600 = 45\cdot6u \times 4\cdot18(T - 290)$$

or
$$T = 290 + 8\cdot391/u \text{ K}$$

$$\theta_1 = (350 - 290) = 60 \text{ K}, \quad \theta_2 = (350 - 290 - 8\cdot391/u) = (60 - 8\cdot391/u)$$

and
$$\theta_m = (60 - 60 + 8\cdot391/u)/\ln[60/(60 - 8\cdot391/u)]$$
$$= 8\cdot391/\{u \ln[60u/(60u - 8\cdot391)]\}$$

In equation 7.9:

$$1600 = [u^{0\cdot8}/(0\cdot190 + 0\cdot644u^{0\cdot8})] \times 20\cdot74 \times 8\cdot391/\{u \ln[60u/(60u - 8\cdot391)]\}$$

or
$$1/\{u^{0\cdot2}(0\cdot190 + 0\cdot644u^{0\cdot8})\ln[60u/(60u - 8\cdot391)]\} = 9\cdot194$$

The left-hand side of this equation is plotted against u in Figure 7e and the function equals $9\cdot194$ when $\underline{u = 2\cdot06 \text{ m/s}}$.

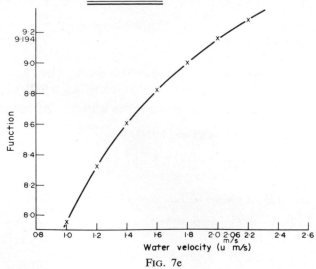

FIG. 7e

Problem 7.55

Derive an expression for the radiant heat transfer rate per unit area between two large parallel planes of emissivities e_1 and e_2 and at absolute temperatures T_1 and T_2 respectively.

Two such planes are situated 2·5 mm apart in air; one has an emissivity of 0·1 and is at a temperature of 350 K, and the other has an emissivity of 0·05 and is at a temperature of 300 K. Calculate the percentage change in the total heat transfer rate by coating the first surface so as to reduce its emissivity to 0·025.

$$\text{Stefan–Boltzmann constant} = 5\cdot67 \times 10^{-8}\ \text{W/m}^2\,\text{K}^4$$

$$\text{Thermal conductivity of air} = 0\cdot026\ \text{W/m K}$$

Solution

The theoretical derivation is laid out in full in Section 7.5.4 from which the heat transfer by radiation is given by equation 7.89:

$$q_r = [(e_1 e_2 \sigma)/(e_1 + e_2 - e_1 e_2)](T_1^4 - T_2^4)$$

For *conduction* between the two planes, in equation 7.11:

$$q_c = kA(T_1 - T_2)/x$$
$$= 0\cdot026 \times 1\cdot0(350 - 300)/0\cdot0025 = 520\ \text{W/m}^2$$

For radiation between the two planes, in equation 7.89:

$$q_r = [(e_1 e_2 \sigma)/(e_1 + e_2 - e_1 e_2)](T_1^4 - T_2^4)$$
$$= [(0\cdot1 \times 0\cdot05 \times 5\cdot67 \times 10^{-8})/(0\cdot1 + 0\cdot05 - 0\cdot1 \times 0\cdot05)](350^4 - 300^4)$$
$$= 13\cdot5\ \text{W/m}^2$$

Thus neglecting any convection in the very narrow space, the total heat transferred is 533·5 W/m². When $e_1 = 0\cdot025$, the heat transfer by radiation becomes:

$$q_r = [(0\cdot025 \times 0\cdot05 \times 5\cdot67 \times 10^{-8})/(0\cdot025 + 0\cdot05 - 0\cdot025 \times 0\cdot05)](350^4 - 300^4)$$
$$= 6\cdot64\ \text{W/m}^2$$

and
$$(q_r + q_c) = 526\cdot64\ \text{W/m}^2$$

Thus although the heat transferred by radiation is reduced to $(100 \times 6\cdot64)/13\cdot5 = 49\cdot2\%$ of its initial value, the total heat transferred is reduced to $(100 \times 526\cdot64)/533\cdot5 = \underline{\underline{98\cdot7\%}}$ of the initial value.

Problem 7.56

Water flows at 2 m/s through a 2·5 m length of a 25 mm diameter tube. If the tube is at 320 K and the water enters and leaves at 293 and 295 K respectively, what is the value of the heat transfer coefficient? How would the outlet temperature change if the velocity were increased by 50%?

Solution

The cross-sectional area of 0·025 m tubing $= (\pi/4) \times 0·025^2 = 0·000491 \text{ m}^2$.
Therefore volume flow of water $= (2 \times 0·000491) = 0·000982 \text{ m}^3/\text{s}$
and mass flow of water $= (1000 \times 0·000982) = 0·982 \text{ kg/s}$.

$\therefore$ Heat load, $Q = 0·982 \times 4·18(295 - 293) = 8·21 \text{ kW}$

Surface area of 0·025 m tubing $= (\pi \times 0·025 \times 1·0) = 0·0785 \text{ m}^2/\text{m}$

or $A = (0·0785 \times 2·5) = 0·196 \text{ m}^2$

$$\theta_1 = (320 - 293) = 27 \text{ K}, \quad \theta_2 = (320 - 295) = 25 \text{ K}$$

and $\theta_m = (27 - 25)/\ln(27/25) = 25·98 \text{ say } 26 \text{ K}$

In equation 7.9: $U = 8·21/(0·196 \times 26) = \underline{1·612 \text{ kW/m}^2 \text{ K}}$

An estimate may be made of the inside film coefficient from equation 7.59, where
T, the mean water temperature is 294 K.

Thus $h_i = 1063(1 + 0·00293 \times 294)2·0^{0·8}/0·025^{0·2}$

$= (1063 \times 1·861 \times 1·741/0·478) = 7205 \text{ W/m}^2 \text{ K or } 7·21 \text{ kW/m}^2 \text{ K}$

The scale resistance is therefore given by:

$$1/1·612 = 1/7·21 + R \quad \text{or} \quad R = 0·482 \text{ m}^2 \text{ K/kW}$$

With a water velocity of $(2·0 \times 150/100) = 3·0 \text{ m/s}$, assuming a mean water temperature of 300 K:

$$h_i = 1063(1 + 0·00293 \times 300)3·0^{0·8}/0·025^{0·2}$$

$$= (1063 \times 1·879 \times 2·408/0·478) = 10063 \text{ or } 10·1 \text{ kW/m}^2 \text{ K}$$

$\therefore$ $1/U = (0·482 + 1/10·1)$

and $U = 1·72 \text{ kW/m}^2 \text{ K}$

For an outlet water temperature of T K, $\theta_1 = 27 \text{ K}, \quad \theta_2 = (320 - T)$
and, taking an arithmetic mean, $\theta_m = 0·5(27 + 320 - T)$
$= 173·5 - 0·5T$

The mass flow of water $= (0·982 \times 150)/100 = 1·473 \text{ kg/s}$,

and the heat load, $Q = 1·473 \times 4·18(T - 293) = (6·157T - 1804) \text{ kW}$

and therefore $(6·157T - 1804) = [1·72 \times 0·196(173·5 - 0·5T)]$

from which $T = \underline{294·4 \text{ K}}$

The use of 300 K as a mean water temperature has a minimal effect on the result and
recalculation is not necessary.

Problem 7.57

A liquid hydrocarbon is fed at 295 K to a heat exchanger consisting of a 25 mm
diameter tube heated on the outside by condensing steam at atmospheric pressure.

The flowrate of hydrocarbon is measured by means of a 19 mm orifice fitted to the 25 mm feed pipe. The reading on a differential manometer containing hydrocarbon-over-water is 450 mm and the coefficient of discharge of the meter is 0·6.

Calculate the initial rate of rise of temperature (K/s) of the hydrocarbon as it enters the heat exchanger.

Outside film coefficient = 6·0 kW/m² K

Inside film coefficient *h* is given by:

$$hd/k = 0·023(\mu d\rho/\mu)^{0·8}(C\mu/k)^{0·4}$$

u = linear velocity of hydrocarbon (m/s)
d = tube diameter (m)
ρ = liquid density (800 kg/m³)
μ = liquid viscosity (9×10^{-4} N s/m²)
C = specific heat of liquid ($1·7 \times 10^3$ J/kg K)
k = thermal conductivity of liquid (0·17 W/m K)

Solution

The effective specific gravity of the liquid = 0·2 or an effective density of 200 kg/m³.
The pressure difference across the orifice = 450 mm water

or $(450 \times 800/200) = 1800$ mm hydrocarbon

i.e. $h = 1·80$ m

The area of the orifice = $(\pi/4) \times 0·019^2 = 2·835 \times 10^{-4}$ m².
In equation 5.19: $G = 0·6 \times 2·835 \times 10^{-4} \times 800\sqrt{(2 \times 9·81 \times 1·80)}$
$\quad = 0·136\sqrt{(35·3)} = 0·808$ kg/s
The volume flow = $(0·808/800) = 0·00101$ m³/s.
Cross area of a 0·025 m diameter pipe = $(\pi/4) \times 0·025^2 = 0·000491$ m²
and hence the velocity, $u = 0·00101^4/0·000491) = 2·06$ m/s.
The inside film coefficient is given by:

$$(h_i \times 0·025/0·17) = 0·023(2·06 \times 0·025 \times 800/9 \times 10^{-4})^{0·2}(1·7 \times 10^3 \times 9 \times 10^{-4}/0·17)^{0·4}$$
$$h_i = 0·1564(4·58 \times 10^4)^{0·8}(9·0)^{0·4}$$
$$= 2016 \text{ W/m}^2 \text{ K} \quad \text{or} \quad 2·02 \text{ kW/m}^2 \text{ K}$$

Neglecting scale and wall resistances:

$$1/U = 1/6·0 + 1/2·02 \quad \text{and} \quad U = 1·511 \text{ kW/m}^2 \text{ K}$$

For steam at atmospheric pressure, the temperature = 373 K and at the inlet the temperature driving force = $(373 - 295) = 78$ K.
The heat flux is thus $(1·511 \times 78) = 117·9$ kW/m².
For a small length of tube, say 0·001 m, the area for heat transfer = $(\pi \times 0·025 \times 0·001) = 7·854 \times 10^{-5}$ m²
and the heat transferred = $(117·9 \times 7·854 \times 10^{-3} \times 1000) = 9·27$ W.
In 0·001 m tube, mass of material = $(0·000491 \times 0·001 \times 800) = 3·93 \times 10^{-4}$ kg
and hence temperature rise $\quad = [9·27/(3·93 \times 10^{-4} \times 1·7 \times 10^3)]$
$$= 13·9 \text{ K/s}$$

Problem 7.58

Water passes at 1·2 m/s through a series of 25 mm diameter tubes 5 m long maintained at 320 K. If the inlet temperature is 290 K, at what temperature would you expect it to leave?

Solution

Assume an outlet water temperature of T K. The mean water temperature is therefore $= 0·5(T + 290) = (0·5T + 145)$ K.

The coefficient may be calculated from equation 7.59:

$$h = 1063(1 + 0·00293T)u^{0·8}/d^{0·2}$$
$$= 1063[1 + 0·00293(0·5T + 145)]1·2^{0·8}/0·025^{0·2}$$
$$= (3665 + 3·77T) \text{ W/m}^2 \text{ K}$$

The area for heat transfer $= (\pi \times 0·025 \times 5·0) = 0·393$ m²
and the heat load, $Q = [1·2 \times (\pi/4) \times 0·025^2 \times 1000 \times 4·18 \times 10^3(T - 290)]$
$$= (2462T - 714,045) \text{ W}$$

Therefore neglecting any scale resistance,

$$(2462T - 714,045) = (3665 + 3·77T) \times 0·393[320 - (0·5T + 145)]$$

from which $\qquad\qquad T^2 + 3937T - 1,300,300 = 0$

and $\qquad\qquad\qquad \underline{\underline{T = 306 \text{ K}}}$

[An alternative approach is as follows:

The heat transferred per unit time in length of pipe, $dL = h \times \pi \times 0·025 \, dL(320 - T_k)$ W where T_k is the water temperature at L m from the inlet.

The rate of increase in the heat content of the water

$$= (\pi/4) \times 0·025^2 \times 1·2 \times 1000 \times 4·18 \times 10^3 \, dT$$
$$= 2462 \, dT$$

The outlet temperature T' is then given by:

$$\int_{290}^{T'} \frac{dT}{(320 - T)} = 0·0000319h \int_0^5 dL$$

∴ $\qquad\qquad \ln(320 - T') = \ln 30 - 0·0001595h$
$$= 3·401 - 0·0001595h$$

At a mean temperature of say 300 K, in equation 7.59:

$$h = 1063(1 + 0·00293 \times 300)1·2^{0·8}/0·025^{0·2} = 4833 \text{ W/m}^2 \text{ K}$$

∴ $\qquad \ln(320 - T') = 3·401 - (0·0001595 \times 4833)$

and $\qquad\qquad \underline{\underline{T' = 306·1 \text{ K}}}$

Problem 7.59

Heat is transferred from one fluid stream to a second fluid across a heat transfer surface. If the film coefficients for the two fluids are, respectively, 1·0 and 1·5 kW/m² K, the metal is 6 mm thick (thermal conductivity 20 W/m K) and the scale coefficient is equivalent to 850 W/m² K, what is the overall heat transfer coefficient?

Solution

From equation 7.130:

$$1/U = 1/h_o + x_w/k_w + R + 1/h_i$$
$$= 1/1000 + 0·006/20 + 1/850 + 1/1500$$
$$= 0·001 + 0·00030 + 0·00118 + 0·00067 = 0·00315$$

∴ $\qquad U = 317·5$ W/m² K or $\underline{\underline{0·318 \text{ kW/m}^2 \text{ K}}}$

Problem 7.60

A pipe of outer diameter 50 mm carries hot fluid at 1100 K. It is covered with a 50 mm layer of insulation of thermal conductivity 0·17 W/m K. Would it be feasible to use magnesia insulation, which will not stand temperatures above 615 K and has a thermal conductivity of 0·09 W/m K for an additional layer thick enough to reduce the outer surface temperature to 370 K in surroundings at 280 K? Take the surface coefficient of transfer by radiation and convection as 10 W/m² K.

Solution

The solution is presented as Problem 7.8.

Problem 7.61

A jacketed reaction vessel containing 0·25 m³ of liquid of specific gravity 0·9 and specific heat 3·3 kJ/kg K is heated by means of steam fed to a jacket on the walls. The contents of the tank are agitated by a stirrer running at 3 Hz. The heat transfer area is 2·5 m² and the steam temperature is 380 K. The outside film heat transfer coefficient is 1·7 kW/m K and the 10 mm thick wall of the tank has a thermal conductivity of 6·0 W/m K. The inside film coefficient was found to be 1·1 kW/m² K for a stirrer speed of 1·5 Hz and to be proportional to the two thirds power of the speed of rotation. Neglecting heat losses and the heat capacity of the tank, how long will it take to raise the temperature of the liquid from 295 to 375 K?

Solution

For a stirrer speed of $1 \cdot 5$ Hz, $h_i = 1 \cdot 1$ kW/m² K.

$\therefore$
$$1 \cdot 1 = K \, 1 \cdot 5^{0 \cdot 67}$$

and
$$K = 0 \cdot 838$$

Thus at a stirrer speed of 3 Hz, $h_i = (0 \cdot 838 \times 3 \cdot 0^{0 \cdot 67}) = 1 \cdot 75$ kW/m² K.
The overall coefficient is given by equation 7.130:

$$1/U = 1/1750 + 0 \cdot 010/6 \cdot 0 + 1/1700 + 0 \cdot 00283$$

and
$$U = 353 \cdot 8 \text{ W/m}^2 \text{ K} \quad \text{neglecting scale resistances.}$$

The time for heating the liquid is given by equation 7.138:

$$\ln[(T_s - T_1)/(T_s - T_2)] = UAt/mC_p$$

In this case, $m = (0 \cdot 25 \times 900) = 225$ kg and $C_p = 3300$ J/kg K.

$\therefore$
$$\ln[(380 - 295)/(380 - 375)] = 353 \cdot 8 \times 2 \cdot 5t/(225 \times 3300)$$
$$2 \cdot 833 = 0 \cdot 00119t$$

and
$$t = \underline{\underline{2381 \text{ s}}} \quad (\approx 40 \text{ min})$$

Problem 7.62

By dimensional analysis, derive a relationship for the heat transfer coefficient h for natural convection between a surface and a fluid on the assumption that the coefficient is a function of the following variables:

k = thermal conductivity of the fluid,
C = specific heat of the fluid,
ρ = density of the fluid,
μ = viscosity of the fluid,
βg = the product of the acceleration due to gravity and the coefficient of cubical expansion of the fluid,
l = a characteristic dimension of the surface, and
T = the temperature difference between the fluid and the surface.

Indicate why each of the above quantities would be expected to influence the heat transfer coefficient and explain how the orientation of the surface affects the process.

Under what conditions is heat transfer by convection important in chemical engineering?

Solution

Suppose that the heat transfer coefficient h can be expressed as a product of porous of the variables.

Let $h = K(k^a C^b \rho^c \mu^d (\beta g)^e l^f T^g)$ where K is a constant.
The dimensions of each variable in terms of **M, L, T, Q,** and $\boldsymbol{\theta}$ are:

$$\begin{aligned}
\text{heat transfer coefficient,} \quad & h = \mathbf{Q/L^2 T\theta} \\
\text{thermal conductivity,} \quad & k = \mathbf{Q/LT\theta} \\
\text{specific heat,} \quad & C = \mathbf{Q/M\theta} \\
\text{viscosity,} \quad & \mu = \mathbf{M/LT} \\
\text{density,} \quad & \rho = \mathbf{M/L^3} \\
\text{the product,} \quad & \beta g = \mathbf{L/T^2 \theta^{-1}} \\
\text{length,} \quad & l = \mathbf{L} \\
\text{temperature difference,} \quad & T = \boldsymbol{\theta}.
\end{aligned}$$

Equating indices:

$$\begin{aligned}
\mathbf{M:} \quad & 0 = -b + c + d \\
\mathbf{L:} \quad & -2 = -a - 3c - d + e + f \\
\mathbf{T:} \quad & -1 = -a - d - 2e \\
\mathbf{Q:} \quad & 1 = a + b \\
\boldsymbol{\theta:} \quad & -1 = -a - b - e + g
\end{aligned}$$

Solving in terms of b and c:

$$a = 1 - b, \quad d = b - c, \quad e = c/2, \quad f = 3c/2 - 1, \quad g = c/2$$

and hence

$$h = K \left(\frac{k}{k^b} C^b \rho^c \frac{\mu^b}{\mu^c} (\beta g)^{c/2} \frac{l^{3c/2}}{l} T^{c/2} \right)$$

$$= K \left(\frac{k}{l} \right) \left(\frac{C\mu}{k} \right)^b \left(\frac{l^{3/2} \rho (\beta g)^{1/2} T^{1/2}}{\mu} \right)^c$$

or

$$\frac{hl}{k} = K \left(\frac{C\mu}{k} \right)^b \left(\frac{l^3 \rho^2 \beta g T}{\mu^2} \right)^{c/2}$$

where $(C\mu/k)$ is the Prandtl number and $(l^3 \rho^2 \beta g T/\mu^2)$ the Grashof number. A full discussion of the significance of this result and the importance of free convection is presented in Section 7.4.7.

Problem 7.63

A shell-and-tube heat exchanger is used for preheating the feed to an evaporator. The liquid of specific heat 4·0 kJ/kg K and specific gravity 1·1 passes through the inside of tubes and is heated by steam condensing at 395 K on the outside. The exchanger heats liquid at 295 K to an outlet temperature of 375 K when the flowrate is $1·75 \times 10^{-4}$ m³/s and to 370 K when the flowrate is $3·25 \times 10^{-4}$ m³/s. What is the heat transfer area, and the value of the overall heat transfer coefficient when the flow is $1·75 \times 10^{-4}$ m³/s?

Assume that the film heat transfer coefficient for the liquid in the tubes is proportional to the 0·8 power of the velocity, that the transfer coefficient for the condensing steam remains constant at 3·4 kW/m² K, and that the resistance of the tube wall and scale can be neglected.

Solution

For a flow of $1.75 \times 10^{-4}\ m^3/s$

The density $= 1100\ kg/m^3$
and hence the mass flow $= (1.75 \times 10^{-4} \times 1100) = 0.1925\ kg/s$.
The heat load $= 0.1925 \times 4.0(373 - 295) = 61.6\ kW$

$$\theta_1 = (395 - 295) = 100\ K, \quad \theta_2 = (395 - 375) = 20\ K$$

and in equation 7.10:

$$\theta_m = (100 - 20)/\ln(100/20) = 49.7\ K$$

Thus, in equation 7.9

$$U_1 A = (61.6/49.7) = 1.239\ kW/K$$

For a flow of $3.25 \times 10^{-4}\ m^3/s$

The mass flow $= (3.25 \times 10^{-4} \times 1100) = 0.3575\ kg/s$
and the heat load $= 0.3575 \times 4.0(370 - 295) = 107.3\ kW$

$$\theta_1 = (395 - 295) = 100\ K, \quad \theta_2 = (395 - 370) = 25\ K$$

and in equation 7.10:

$$\theta_m = (100 - 25)/\ln(100/25) = 54.1\ K$$

Thus in equation 7.9:

$$U_2 A = (107.3/54.1) = 1.983\ kW/K$$
$$\therefore \qquad U_2/U_1 = (1.983/1.239) = 1.60$$

The velocity in the tubes is proportional to the volumetric flowrate, $v\ cm^3/s$ and hence

$$h_i \propto v^{0.8} \quad or \quad h_i = K\,v^{0.8}$$

Neglecting scale and wall resistances:

$$1/U = 1/h_o + 1/h_i$$
$$= 1/3.4 + 1/K\,v^{0.8}$$

and $\qquad U = 3.4Kv^{0.8}/(3.4 + Kv^{0.8})$

$\therefore \qquad U_1 = (3.4K \times 175^{0.8})/(3.4 + K \times 175^{0.8}) = 211.8K/(3.4 + 62.3K)$

and $\qquad U_2 = (3.4K \times 325^{0.8})/(3.4 + K \times 325^{0.8}) = 347.5K/(3.4 + 102.2K)$

$\therefore \qquad [347.5K/(3.4 + 102.2K)]/[211.18K/(3.4 + 62.3K)] = 1.60$

$$\therefore \qquad K = 0.00228$$

$\therefore \qquad U_1 = (3.4 \times 0.00228 \times 175^{0.8})/(3.4 + 0.00228 \times 175^{0.8}) = \underline{\underline{0.136\ kW/m^2\ K}}$

The heat transfer area, $A = 1.239/0.136 = \underline{\underline{9.09\ m^2}}$.

SECTION 8

MASS TRANSFER

Problem 8.1

Ammonia gas is diffusing at a constant rate through a layer of stagnant air 1 mm thick. Conditions are fixed so that the gas contains 50% by volume of ammonia at one boundary of the stagnant layer. The ammonia diffusing to the other boundary is quickly absorbed and the concentration is negligible at that plane. The temperature is 295 K and the pressure atmospheric, and under these conditions the diffusivity of ammonia in air is 0·18 cm²/s. Calculate the rate of diffusion of ammonia through the layer.

Solution

Let subscripts 1 and 2 refer to the two sides of the stagnant layer and subscripts A and B refer to ammonia and air respectively.

Equation 8.23 gives the rate of diffusion through a stagnant layer as:

$$N_A = -(D/\mathbf{R}Tx)(P/P_{BM})(P_{A2} - P_{A1})$$

In this problem: $x = 1 \text{ mm} = 0·001 \text{ m}$
$D = 0·18 \text{ cm}^2/\text{s} = 0·18 \times 10^{-4} \text{ m}^2/\text{s}$
$\mathbf{R} = 8·314 \text{ kJ/kmol K}$
$T = 295 \text{ K}$
$P = 101·3 \text{ kN/m}^2$
$P_{A1} = 0·50 \times 101·3 = 50·65 \text{ kN/m}^2$
$P_{A2} = 0$
$P_{B1} = 101·3 - 50·65 = 50·65 \text{ kN/m}^2$
$P_{B2} = 101·3 - 0 = 101·3 \text{ kN/m}^2$
$P_{BM} = (101·3 - 50·65)/\ln(101·3/50·65) = 73·07 \text{ kN/m}^2$

Hence by substitution:

$$N_A = -(0·18 \times 10^{-4}/8·314 \times 295 \times 10^{-3})(101·3/73·07)(0 - 50·65)$$
$$= 5·15 \times 10^{-4} \text{ kmol/m}^2 \text{ s}$$

Problem 8.2

A simple rectifying column consists of a tube, arranged vertically and supplied at the bottom with a mixture of benzene and toluene as vapour. At the top, a condenser returns some of the product as a reflux which flows in a thin film down the inner wall of the tube. The tube is insulated and heat losses can be neglected. At one point in the column, the vapour contains 70 mol% benzene and the adjacent liquid reflux contains 59 mol% benzene. The temperature at this point is 365 K. Assuming the diffusional resistance to vapour transfer to be equivalent to the diffusional resistance of a stagnant vapour layer 0·2 mm thick, calculate the rate of interchange of benzene and toluene between vapour and liquid. The molar latent heats of the two materials can be taken as equal. The vapour pressure of toluene at 365 K is 54·0 kN/m² and the diffusivity of the vapours is 0·051 cm²/s.

Solution

Let subscripts 1 and 2 refer to the liquid surface and vapour side of the stagnant layer respectively and let subscripts B and T refer to benzene and toluene.

If the latent heats are equal and there are no heat losses, there is no change of phase across the stagnant layer.

This is an example of equimolecular counter diffusion and equation 8.16 applies:

$$N_A = -D(P_{A2} - P_{A1})/\mathbf{R}Tx$$

where x = thickness of the stagnant layer = 0·2 mm = 0·0002 m.

As the vapour pressure of toluene = 54 kN/m², the partial pressure of toluene from Raoult's law = $(1 - 0·59) \times 54 = 22·14$ kN/m² = P_{T1} and

$$P_{T2} = (1 - 0·70) \times 101·3 = 30·39 \text{ kN/m}^2$$

For toluene: $N_T = -(0·051 \times 10^{-4})(30·39 - 22·4)/(8·314 \times 365 \times 0·0002)$

$$= \underline{\underline{-6·93 \times 10^{-5} \text{ kmol/m}^2 \text{ s}}}$$

For benzene: $P_{B1} = 101·3 - 22·14 = 79·16 \text{ kN/m}^2$

$$P_{B2} = 101·3 - 30·39 = 70·91 \text{ kN/m}^2$$

Hence, for benzene: $N_B = -(0·051 \times 10^{-4})(70·91 - 79·16)/(8·314 \times 365 \times 0·0002)$

$$= \underline{\underline{6·93 \times 10^{-5} \text{ kmol/m}^2 \text{ s}}}$$

Thus the rate of interchange of benzene and toluene is equal but opposite in direction.

Problem 8.3

By what percentage would the rate of absorption be increased or decreased by increasing the total pressure from 100 to 200 kN/m² in the following cases?

(a) The absorption of ammonia from a mixture of ammonia and air containing 10% of ammonia by volume, using pure water as solvent. Assume that all the resistance to mass transfer lies within the gas phase.

(b) The same conditions as (a) but the absorbing solution exerts a partial vapour pressure of ammonia of 5 kN/m².

The diffusivity can be assumed to be inversely proportional to the absolute pressure.

Solution

(a) Use is made of equation 8.23 to find the rates of diffusion for the two pressures:

$$N_A = -(D/\mathbf{R}Tx)(P/P_{BM})(P_{A2} - P_{A1})$$

Let subscripts 1 and 2 refer to water and air side of the layer respectively and subscripts A and B refer to ammonia and air.

$$P_{A2} = 0.10 \times 100 = 10 \text{ kN/m}^2 \quad \text{and} \quad P_{A1} = 0 \text{ kN/m}^2$$
$$P_{B2} = 100 - 10 = 90 \text{ kN/m}^2 \quad \text{and} \quad P_{B1} = 100 \text{ kN/m}^2$$
$$P_{BM} = (100 - 90)/\ln(100/90) = 94.91 \text{ kN/m}^2$$
$$P/P_{BM} = 100/94.91 = 1.054$$

Hence
$$N_A = -(D/\mathbf{R}Tx)1.054(10 - 0)$$
$$= -10.54D/\mathbf{R}Tx$$

If the pressure is doubled to 200 kN/m², the diffusivity is halved to $0.5D$ (from equation 8.44):

$$P_{A2} = 0.1 \times 200 = 20 \text{ kN/m}^2 \quad \text{and} \quad P_{A1} = 0 \text{ kN/m}^2$$
$$P_{B2} = 200 - 20 = 180 \text{ kN/m}^2 \quad \text{and} \quad P_{B1} = 200 \text{ kN/m}^2$$
$$P_{BM} = (200 - 180)/\ln(200/180) = 189.82 \text{ kN/m}^2$$
$$P/P_{BM} = 200/189.82 = 1.054 \text{ i.e. unchanged}$$

Hence
$$N_A = -(0.5D/\mathbf{R}Tx)1.054(20 - 0)$$
$$= -10.54D/\mathbf{R}Tx \text{ i.e. the rate is unchanged}$$

(b) If the absorbing solution now exerts a partial vapour pressure of ammonia of 5 kN/m², then at a total pressure of 100 kN/m²:

$$P_{A2} = 10 \text{ kN/m}^2 \quad \text{and} \quad P_{A1} = 5 \text{ kN/m}^2$$
$$P_{B2} = 90 \text{ kN/m}^2 \quad \text{and} \quad P_{B1} = 95 \text{ kN/m}^2$$
$$P_{BM} = (95 - 90)/\ln(95/90) = 92.48 \text{ kN/m}^2$$
$$P/P_{BM} = 100/92.48 = 1.081$$
$$N_A = -(D/\mathbf{R}Tx) \times 1.081(10 - 5) = -5.406D/\mathbf{R}Tx$$

At 200 kN/m^2, the diffusivity $= 0.5D$:

$$P_{A2} = 20 \text{ kN/m}^2 \quad \text{and} \quad P_{A1} = 5 \text{ kN/m}^2$$
$$P_{B2} = 180 \text{ kN/m}^2 \quad \text{and} \quad P_{B1} = 195 \text{ kN/m}^2$$
$$P_{BM} = (195 - 180)/\ln(195/180) = 187.4 \text{ kN/m}^2$$
$$P/P_{BM} = 1.067$$
$$N_A = -(0.5D/\mathbf{R}Tx)1.067(20 - 5)$$
$$= -8.0D/\mathbf{R}Tx$$

The rate of diffusion has been increased by $(8 - 5.406)/5.406$ which is equal to 48%.

Problem 8.4

In the Danckwerts model of mass transfer it is assumed that the fractional rate of surface renewal s is constant and independent of surface age. Under such conditions the expression for the surface age distribution function ϕ is $\phi = se^{-st}$.

If the fractional rate of surface renewal were proportional to surface age (say $s = bt$, where b is a constant), show that the surface age distribution function would then assume the form:

$$\phi = (2b/\pi)^{1/2} e^{-bt^2/2}$$

Solution

From equation 8.78: $\qquad\qquad\qquad f'(t) + sf(t) = 0$
In this problem, $s = bt$ and:

$$e^{bt^2/2}f(t) = \text{constant} = k$$
$$\therefore \qquad f(t) = k\,e^{-bt^2/2}$$

The total area of surface considered is unity:

$$\therefore \qquad \int_0^\infty f(t)\,\mathrm{d}t = 1$$

$$\therefore \qquad \int_0^\infty k\,e^{-bt^2/2}\,\mathrm{d}t = 1$$

and by substitution as in equation 8.81:

$$k(\pi/2b)^{0.5} = 1$$
$$k = (2b/\pi)^{0.5}$$

and $\qquad\qquad\qquad f(t) = (2b/\pi)^{\frac{1}{2}} e^{-bt^2/2}$

Problem 8.5

By consideration of the appropriate element of a sphere show that the general equation for molecular diffusion in a stationary medium and in the absence of a chemical reaction is:

$$\frac{\partial C}{\partial t} = D\left(\frac{\partial^2 C}{\partial r^2} + \frac{1}{r^2}\frac{\partial^2 C}{\partial \beta^2} + \frac{1}{r^2 \sin^2 \beta}\frac{\partial^2 C}{\partial \phi^2} + \frac{2\partial C}{r \partial r} + \frac{\cot \beta}{r^2}\frac{\partial C}{\partial \beta}\right)$$

where C is the concentration of the diffusing substance, D the molecular diffusivity, t the time, and r, β, ϕ are spherical polar coordinates, β being the latitude angle.

Solution

The basic equation for unsteady state mass transfer is equation 8.50, i.e.:

$$\frac{\partial C}{\partial t} = D\left[\left(\frac{\partial^2 C}{\partial x^2}\right)_{yz} + \left(\frac{\partial^2 C}{\partial y^2}\right)_{zx} + \left(\frac{\partial^2 C}{\partial z^2}\right)_{xy}\right] \tag{1}$$

This equation may be transformed into other systems of orthogonal coordinates, the most useful being the spherical polar system. The reader is referred to Carslaw and Jaeger, *Conduction of Heat in Solids*, for details of the transformation. When the operation is performed:

$$x = r \sin \beta \cos \phi$$
$$y = r \sin \beta \sin \phi$$
$$z = r \cos \beta$$

and the equation for C becomes:

$$\frac{\partial C}{\partial t} = \frac{D}{r^2}\left[\frac{\partial}{\partial r}\left(r^2 \frac{\partial C}{\partial r}\right) + \frac{1}{\sin \beta}\frac{\partial}{\partial \beta}\left(\sin \beta \frac{\partial C}{\partial \beta}\right) + \frac{1}{\sin^2 \beta}\frac{\partial^2 C}{\partial \phi^2}\right] \tag{2}$$

which may be written as:

$$\frac{\partial C}{\partial t} = D\left[\frac{\partial^2 C}{\partial r^2} + \frac{2}{r}\frac{\partial C}{\partial r} + \frac{1}{r^2}\frac{\partial}{\partial \mu}\left((1 - \mu^2)\frac{\partial C}{\partial \mu}\right) + \frac{1}{r^2(1 - \mu^2)}\frac{\partial^2 C}{\partial \phi^2}\right] \tag{3}$$

where $\mu = \cos \beta$. $\tag{4}$

In this problem $\partial C/\partial t$ is given by:

$$\frac{\partial C}{\partial t} = D\left(\frac{\partial^2 C}{\partial r^2} + \frac{1}{r^2}\frac{\partial^2 C}{\partial \beta^2} + \frac{1}{r^2 \sin^2 \beta}\frac{\partial^2 C}{\partial \phi^2} + \frac{2}{r}\frac{\partial C}{\partial r} + \frac{\cot \beta}{r^2}\frac{\partial C}{\partial \beta}\right) \tag{5}$$

Comparing equations (3) and (5) is necessary to prove that:

$$\frac{1}{r^2}\frac{\partial}{\partial \mu}\left((1 - \mu^2)\frac{\partial C}{\partial \mu}\right) + \frac{1}{r^2(1 - \mu^2)}\frac{\partial^2 C}{\partial \phi^2} = \frac{1}{r^2}\frac{\partial^2 C}{\partial \beta^2} + \frac{1}{r^2 \sin^2 \beta}\frac{\partial^2 C}{\partial \phi^2} + \frac{\cot \beta}{r^2}\frac{\partial C}{\partial \beta}$$

Now $\mu = \cos \beta$, $1 - \mu^2 = 1 - \cos^2 \beta = \sin^2 \beta$

$$\therefore \qquad \frac{1}{r^2(1 - \mu^2)}\frac{\partial^2 C}{\partial \phi^2} = \frac{1}{r^2 \sin^2 \beta}\frac{\partial^2 C}{\partial \phi^2}$$

It now becomes necessary to prove that:

$$\frac{1}{r^2}\frac{\partial^2 C}{\partial \beta^2} + \frac{\cot \beta}{r^2}\frac{\partial C}{\partial \beta} = \frac{1}{r^2}\frac{\partial}{\partial \mu}\left((1-\mu^2)\frac{\partial C}{\partial \mu}\right) \qquad (6)$$

From (4),
$$\mu = \cos \beta$$

∴
$$\partial\mu/\partial\beta = -\sin \beta \qquad (7)$$

and
$$\partial^2\mu/\partial\beta^2 = -\cos \beta \qquad (8)$$

$$\frac{1}{r^2}\frac{\partial}{\partial\mu}\left((1-\mu^2)\frac{\partial C}{\partial\mu}\right) = \frac{1}{r^2}\frac{\partial}{\partial\beta}\left((1-\mu^2)\frac{\partial C}{\partial\beta}\frac{\partial\beta}{\partial\mu}\right)\frac{\partial\beta}{\partial\mu}$$

Substituting from (4) for μ; from (7) for $\partial\beta/\partial\mu$ gives:

$$= \frac{1}{r^2}\frac{\partial}{\partial\beta}\left((1-\cos^2\beta)\frac{\partial C}{\partial\beta}\frac{1}{-\sin\beta}\right)\frac{1}{-\sin\beta}$$

$$= \frac{1}{r^2}\frac{\partial}{\partial\beta}\left(-\sin\beta\frac{\partial C}{\partial\beta}\right)\frac{1}{-\sin\beta}$$

$$= \frac{1}{r^2}\left[-\sin\beta\frac{\partial^2 C}{\partial\beta^2} + \frac{\partial C}{\partial\beta}(-\cos\beta)\right]\frac{1}{-\sin\beta}$$

$$= \frac{1}{r^2}\frac{\partial^2 C}{\partial\beta^2} + \frac{\cot\beta}{r^2}\frac{\partial C}{\partial\beta}$$

which finally proves the required relationship.

Problem 8.6

Prove that for equimolecular counter diffusion from a sphere to a surrounding stationary, infinite medium, the Sherwood number based on the diameter of the sphere is equal to 2.

Solution

If r_o = radius of the sphere and r is the distance from the centre of the sphere to a point in the surrounding medium, then by analogy with heat transfer:

$$N = kA\Delta C = DA\Delta C/(r_o - r)$$

or
$$k, \text{ the mass transfer coefficient} = D/(r_o - r)$$

The Sherwood number = kd/D where d is the sphere diameter.

Substituting for k,
$$Sh = Dd/D(r_o - r) = d/(r_o - r)$$

At the sphere surface, $(r_o - r)$ approaches $r_o = d/2$.

Hence
$$Sh = d/(d/2) = \underline{\underline{2}}$$

Problem 8.7

Show that the concentration profile for unsteady-state diffusion into a bounded medium of thickness L, when the concentration at the interface is suddenly raised to a constant value C_i and kept constant at the initial value of C_o at the other boundary is:

$$C = C_o + (C_i - C_o)\left(1 - \frac{z}{L} - \frac{2}{\pi} \sum_{n=0}^{n=\infty} \frac{1}{n} \exp\left(-n^2\pi^2 Dt/L^2\right) \sin nz\pi/L\right)$$

NB—Assume the solution to be the sum of the solution for infinite time (steady-state part) and the solution of a second unsteady-state part; this simplifies the boundary conditions for the second part.

Solution

The system is shown in Fig. 8a.

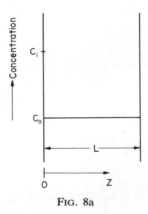

FIG. 8a

The boundary conditions are as follows:

At time $t = 0$ $C = C_o$ $0 < z < L$
        $t > 0$ $C = C_i$ $z = 0$
        $t > 0$ $C = C_o$ $z = L$

Replace C_i by C_i' and C by C' where:

$$C = C' + C_o$$
$$C_i = C_i' + C_o$$

Using these new variables:

At $t = 0$ $C' = 0$ $0 < z < L$
   $t > 0$ $C' = C_i'$ $z = 0$
   $t > 0$ $C' = 0$ $z = L$

The statement of the problem says that the solution of the one dimensional diffusion equation is:

$$C' = (\text{steady state solution}) + \sum_0^\infty \exp(-n^2\pi^2 Dt/L^2) An \sin(n\pi z/L)$$

where the steady state solution $= C'_i - C'_i x/L$.

(A derivation of the analogous equation for heat transfer may be found in *Conduction of Heat in Solids* by H. S. Carslaw and J. C. Jaeger, Oxford, 1960.)

$$An = \frac{2}{L}\int_0^L (\text{Initial concentration profile-steady state}) \sin(n\pi z/L)\, dz$$

$$= \frac{2}{L}\int_0^L [0 + (C'_i x/L) - C'_i] \sin(n\pi z/L)\, dz$$

$$= -2C'_i/n\pi \quad \text{(this proof is given at the end of this problem).}$$

Hence

$$C' = C'_i - C'_i x/L - \frac{2C'_i}{\pi}\sum_{n=0}^\infty \frac{1}{n}\exp(-n^2\pi^2 Dt/L^2)\sin(n\pi z/L)$$

$\therefore$

$$C = C_o + (C_i - C_o)\left[1 - \frac{z}{L} - \frac{2}{\pi}\sum_{n=0}^\infty \frac{1}{n}\exp(-n^2\pi^2 Dt/L^2)\sin(n\pi z/L)\right]$$

A_n is determined as follows:

$$A_n = \frac{2}{L}\int_0^L (C'_i z/L - C'_i)\sin(n\pi z/L)\, dz$$

$$= \frac{2C'_i}{L^2}\int_0^L z\sin(n\pi z/L)\, dz - \frac{2C'_i}{L}\int_0^L \sin(n\pi z/L)\, dz$$

$$= \frac{2C'_i}{L^2}\int_0^L ① - \frac{2C'_i}{L}\int_0^L ②$$

$$\int_0^L ① = \left[-\frac{Lz}{n\pi}\cos\frac{n\pi z}{L}\right]_0^L + \int_0^L \frac{L}{n\pi}\cos\frac{n\pi z}{L}\, dz$$

Put $u = z$, $du = dz$

$$dv = \sin(n\pi z/L)\, dz, \quad v = -\frac{L}{n\pi}\cos\frac{n\pi x}{L}$$

$\therefore$

$$\int_0^L ① = \left(-\frac{Lz}{n\pi}\cos\frac{n\pi z}{L}\right)_0^L + \left(\frac{L^2}{n^2\pi^2}\sin\frac{n\pi z}{L}\right)_0^L$$

$$= -\frac{L^2}{n\pi}\cos n\pi + \frac{L^2}{n^2\pi^2}\sin n\pi = -\frac{L^2}{n\pi}(-1)^n$$

$$\int_0^L ② = \left(-\frac{L}{n\pi} \cos \frac{n\pi z}{L}\right)_0^L = -\frac{L}{n\pi} \cos n\pi + \frac{L}{n\pi}$$

$$= -\frac{L}{n\pi} \cos n\pi + \frac{L}{n\pi}$$

$$= -\frac{L}{n\pi}(-1)^n + \frac{L}{n\pi}$$

$$A_n = \frac{2C_i'}{L^2} ① - \frac{2C_i'}{L} ② = \frac{2C_i'}{L^2}\left(-\frac{L^2}{n\pi}(-1)^n\right) - \frac{2C_i'}{L}\left(-\frac{L}{n\pi}(-1)^n + \frac{L}{n\pi}\right)$$

$$= -\frac{2C_i'}{n\pi}$$

Problem 8.8

Show that under the conditions specified in Problem 8.7 and assuming the Higbie model of surface renewal, the average mass flux at the interface is given by:

$$N_A = (C_i - C_o)D/L \left\{1 + (2L^2/\pi^2 Dt) \sum_{n=1}^{n=\infty}\left[\frac{\pi^2}{6} - \frac{1}{n^2}\exp\left(-n^2\pi^2 Dt/L^2\right)\right]\right\}$$

NB—Use the relation $\sum_{n=1}^{\infty} \frac{1}{n^2} = \pi^2/6$.

Solution

The rate of transference across the phase boundary is given by:

$$N_A = -D(\partial C/\partial z)_{z=0}$$

According to the Higbie model, if the element is exposed for a time t, the average rate of transfer is given by:

$$N_A = \frac{1}{t}\int_0^t -D(\partial C/\partial z)_{z=0}\, dt$$

From Problem 8.7, the concentration C is given by:

$$C = C_o + (C_i - C_o)\left[1 - \frac{z}{L} - \frac{2}{\pi}\sum_{n=0}^{\infty}\frac{1}{n}\exp\left(-n^2\pi^2 Dt/L^2\right)\sin n\pi z/L\right]$$

$$\frac{\partial C}{\partial z} = (C_i - C_o)\left[-\frac{1}{L} - \frac{2}{\pi}\sum_{n=0}^{\infty}\frac{\pi}{L}\exp\left(-n^2\pi^2 Dt/L^2\right)\cos(n\pi z/L)\right]$$

$$\left(\frac{\partial C}{\partial z}\right)_{z=0} = (C_i - C_o)\left[-\frac{1}{L} - \frac{2}{\pi}\sum_{0}^{\infty}\frac{\pi}{L}\exp\left(-n^2\pi^2 Dt/L^2\right)\right]$$

$$N_A = -\frac{D(C_i - C_o)}{t} \int_0^t \left[-\frac{1}{L} - \frac{2}{\pi} \sum_0^\infty \frac{\pi}{L} \exp\left(-n^2\pi^2Dt/L^2\right) \right]$$

$$= -\frac{D(C_i - C_o)}{t} \left[-\frac{t}{L} - \frac{2}{\pi} \sum_0^\infty \frac{\pi}{L} \left(-\frac{L^2}{n^2\pi^2D}\right) \exp\left(-n^2\pi^2Dt/L^2\right) \right]_0^L$$

$$= -\frac{D(C_i - C_o)}{t} \left[-\frac{t}{L} - \frac{2}{\pi} \sum_0^\infty \left(-\frac{L}{n^2\pi D}\right) \exp\left(-n^2\pi^2Dt/L^2\right) \right.$$

$$\left. + \frac{2}{\pi} \sum_0^\infty \frac{\pi}{L} \left(-\frac{L^2}{n^2\pi^2D}\right) \right]$$

$$N_A = \frac{D}{L}(C_i - C_o)\left\{ 1 + \frac{2L^2}{\pi^2Dt} \left[\sum_0^\infty \frac{-1}{n^2} \exp\left(-n^2\pi^2Dt/L^2\right) + \sum_0^\infty \frac{1}{n^2} \right] \right\}$$

Now

$$\sum_0^\infty \frac{-1}{n^2} \exp\left(-n^2\pi^2Dt/L^2\right) + \sum_0^\infty \frac{1}{n^2}$$

$$= \sum_0^1 \frac{-1}{n^2} \exp\left(-n^2\pi^2Dt/L^2\right) + \sum_1^\infty \frac{-1}{n^2} \exp\left(-n^2\pi^2Dt/L^2\right) + \sum_0^1 \frac{1}{n^2} + \sum_1^\infty \frac{1}{n^2}$$

$$= -\exp\left(-\pi^2Dt/L^2\right) + \sum_1^\infty \frac{-1}{n^2} \exp\left(-n^2\pi^2Dt/L^2\right) + 1 + \pi^2/6$$

$$= \sum_1^\infty \left[\frac{\pi^2}{6} - \frac{1}{n^2} \exp\left(-n^2\pi^2Dt/L^2\right) + 1 - \exp\left(-\pi^2Dt/L^2\right) \right]$$

Consider the terms $1 - \exp\left(-\pi^2Dt/L^2\right)$. Dt/L^2 is very small so that $\left(-\pi^2Dt/L^2\right)$ is small and $\exp\left(-\pi^2Dt/L^2\right) \to 1$. Therefore, $1 - \exp\left(-\pi^2Dt/L^2\right)$ is approximately zero and:

$$N_A = \frac{D}{L}(C_i - C_o)\left\{ 1 + \frac{2L^2}{\pi^2Dt} \sum_{n=1}^\infty \left[\frac{\pi^2}{6} - \frac{1}{n^2} \exp\left(-n^2\pi^2Dt/L^2\right) \right] \right\}$$

Problem 8.9

According to the simple penetration theory the instantaneous mass flux, N_A^o is:

$$N_A^o = (C_i - C_o)\left(\frac{D}{\pi t}\right)^{0.5}$$

What is the equivalent expression for the instantaneous heat flux under analogous conditions?

Pure SO_2 is absorbed at 295 K and atmospheric pressure into a laminar water jet. The solubility of SO_2, assumed constant over a small temperature range, is $1.54\,kmol/m^3$ under these conditions and the heat of solution is 28 kJ/kmol.

Calculate the resulting jet surface temperature if the Lewis number is 90. Neglect heat transfer between the water and the gas.

Solution

The heat flux at any time, $f = -k(\partial\theta/\partial x)$ where k is the thermal conductivity, θ the temperature, and x the distance.

The flux satisfies the same differential equation as θ, i.e.:

$$D_H(\partial^2 f/\partial x^2) = (\partial f/\partial t) \quad x > 0, t > 0$$

where D_H = thermal diffusivity = $k/\rho C_p$.

NB—This last equation is analogous to the mass transfer equation 7.32, i.e. $(\partial C/\partial t) = D(\partial^2 C/\partial x^2)$.

The solution of the heat transfer equation with $f = F_o$ (constant) at $x = 0$ when $t > 0$ is:

$$f = F_o \text{ erfc} \frac{x}{2\sqrt{D_H t}}$$

The temperature rise is due to the heat of solution H_s. Heat is liberated at the jet surface at a rate $\mathcal{H}(t) = N_A^o H_A$,

i.e.
$$\mathcal{H}(t) = (C_i - C_o)H_A(D/\pi t)^{0.5}$$

The temperature rise T due to the heat flux $\mathcal{H}(t)$ into the surface is given by:

$$T = \frac{1}{\rho C_p \sqrt{\pi D_H}} \int_0^L \frac{\mathcal{H}(t-\theta)\, d\theta}{\sqrt{\theta}}$$

and
$$T = \frac{(C_i - C_o)H_A\sqrt{D/D_H}}{\rho C_p}$$

The Lewis number = $h/C_p\rho h_D$, which for the same Reynolds number

$$= Pr/Sc = DC_p\rho/k.$$

$$D/D_H = DC_p\rho/k = 90$$

$$C_i - C_o = 1\cdot54 \text{ kmol/m}^3$$

$$H_A = 28 \text{ kJ/kmol}$$

∴
$$T = (1\cdot54 \times 28\sqrt{90})/(1000 \times 4\cdot186)$$

$$= 0\cdot1 \text{ K}$$

Problem 8.10

In a packed column, operating at approximately atmospheric pressure and 295 K, a 10% ammonia–air mixture is scrubbed with water and the concentration is reduced to 0·1%. If the whole of the resistance to mass transfer may be regarded as lying within a thin laminar film on the gas side of the gas–liquid interface, derive from first principles an expression for the rate of absorption at any position in the column. At some intermediate point where the ammonia concentration in the gas phase has been reduced to 5%, the partial pressure of ammonia in equilibrium with the aqueous solution is $660\,N/m^2$ and the transfer rate is $10^{-3}\,kmol/m^2\,s$. What is the thickness of the hypothetical gas film if the diffusivity of ammonia in air is $0.24\,cm^2/s$?

Solution

The equation for the rate of absorption is derived in Section 8.2.1 as equation 8.16:

$$N_A = -(D/\mathbf{R}Tx)(P_{A2} - P_{A1})$$

Let subscripts 1 and 2 refer to the water and air side of the stagnant film and subscripts A and B refer to ammonia and air.

$$P_{A1} = 66.0\,kN/m^2 \quad \text{and} \quad P_{A2} = 0.05 \times 101.3 = 5.065\,kN/m^2$$
$$D = 0.24 \times 10^{-4}\,m^2/s$$
$$\mathbf{R} = 8.314\,kJ/kmol\,K$$
$$T = 295\,K$$
$$N_A = 1 \times 10^{-3}\,kmol/m^2\,s$$

Hence
$$x = -(D/N_A\mathbf{R}T)(P_{A2} - P_{A1})$$
$$= -(0.24 \times 10^{-4}/10^{-3} \times 8.314 \times 295)(66.0 - 5.065)$$
$$= -0.000043\,m$$

The negative sign indicates that the diffusion is taking place in the opposite direction and the thickness of the gas film is <u>0.043 mm</u>.

Problem 8.11

An open bowl, $0.3\,m$ in diameter, contains water at 350 K evaporating into the atmosphere. If the air currents are sufficiently strong to remove the water vapour as it is formed and if the resistance to its mass transfer in air is equivalent to that of a 1 mm layer for conditions of molecular diffusion, what will be the rate of cooling due to evaporation? The water can be considered as well mixed and the water equivalent of the system is equal to 10 kg. The diffusivity of water vapour in air may be taken as $0.20\,cm^2/s$ and the kilogram molecular volume at NTP as $22.4\,m^3$.

Solution

Let subscripts 1 and 2 refer to the water and air side of the stagnant layer and let subscripts A and B refer to water vapour and air.

Equation 8.23 gives the rate of diffusion through a stagnant layer as:

$$N_A = -(D/\mathbf{R}Tx)(P/P_{BM})(P_{A2} - P_{A1})$$

P_{A1} is the vapour pressure of water at 350 K = 41·8 kN/m².
$P_{A2} = 0$ since the air currents remove the vapour as it is formed.
$P_{B1} = 101·3 - 41·8 = 59·5$ kN/m² and $P_{B2} = 101·3$ kN/m².
$P_{BM} = (101·3 - 59·5)/\ln(101·3/59·5) = 78·17$ kN/m².
$P/P_{BM} = 101·3/78·17 = 1·296$.

$$N_A = -(0·2 \times 10^{-4}/8·314 \times 350 \times 10^{-3})1·296(0 - 41·8)$$
$$= 3·72 \times 10^{-4} \text{ kmol/m}^2 \text{ s}$$
$$= 3·72 \times 10^{-4} \times 18 = 6·70 \times 10^{-3} \text{ kg water/m}^2 \text{ s}$$

Area of bowl = $(\pi/4)0·3^2 = 0·0707$ m²
Therefore rate of evaporation = $6·70 \times 10^{-3} \times 0·0707 = 4·74 \times 10^{-4}$ kg/s
Latent heat of vaporisation = 2466 kJ/kg
Specific heat of water = 4·187 kJ/kg K
Rate of heat removal = $4·74 \times 10^{-4} \times 2466$
$$= 1·17 \text{ kW}$$
If the rate of cooling = $d\theta/dt$ K/s,

$$\text{(water equivalent)} \times \text{(specific heat)} \times (d\theta/dt) = 0·0617$$

i.e. $10 \times 4·187 \times (d\theta/dt) = 1·17$

and $d\theta/dt = \underline{\underline{0·028 \text{ K/s}}}$

Problem 8.12

Show by substitution that when a gas of solubility C^+ is absorbed into a stagnant liquid of infinite depth, the concentration at time t and depth x is

$$C^+\text{erfc}\, \frac{x}{2\sqrt{Dt}}$$

Hence, on the basis of the simple penetration theory, show that the rate of absorption in a packed column will be proportional to the square root of the diffusivity.

Solution

The first part of this question is discussed fully in Section 8.4.1 and the required equation is presented as equation 8.58.
In Section 8.4.1 the analysis following equation 8.58 leads to equation 8.60 which expresses the instantaneous rate of mass transfer when the surface element under consideration has an age t, i.e.

$$R = (C_i - C_o)\sqrt{(D/\pi t)}$$

The simple penetration theory assumes that each element is exposed for the same time interval t_e before returning to the bulk solution. The average rate of mass transfer is then given by:

$$N_a = \frac{1}{t_e} \int_0^{t_e} R \, dt = \frac{(C_i - C_o)}{t_e} \int_0^{t_e} (D/\pi t)^{0.5} \, dt$$

$$= 2(C_i - C_o)\sqrt{D/\pi t_e}$$

i.e. rate of absorption is proportional to $\sqrt{D}$.

Problem 8.13

Show that in steady-state diffusion through a film of liquid, accompanied by a first order irreversible reaction, the concentration of solute in the film at depth z below the interface is given by:

$$C = \sinh \frac{\sqrt{\frac{\alpha}{D}}(z_L - z)}{\sinh \sqrt{\frac{\alpha}{D}} z_L} C_i$$

if $C = 0$ at $z = z_L$ and $C = C_i$ at $z = 0$, corresponding to the interface.

Hence show that according to the *film theory* of gas-absorption, the rate of absorption per unit area of interface, N_A is given by:

$$N_A = K_L C_i \frac{\beta}{\tanh \beta}$$

where $\beta = \sqrt{D_\alpha / K_L}$
   D = diffusivity of the solute,
   α = rate constant of the reaction,
   K_L = liquid film mass transfer coefficient for physical absorption,
   C_i = concentration of solute at the interface,
   z = distance normal to the interface and
   z_L = liquid film thickness.

Solution

The basic equation for diffusion through a film of liquid accompanied by a first-order irreversible reaction is presented as equation 8.124:

$$D(d^2 C/dz^2) = \alpha C$$

or $$(d^2 C/dz^2) = a^2 C \tag{1}$$

where $a = \sqrt{\alpha/D}$.

The general solution of (1) is:

$$C = A \cosh az + B \sinh az \tag{2}$$

where A and B are constants.

The boundary conditions are given as:

$$\text{At } z = z_L, \quad C = 0 \tag{3}$$

$$\text{At } z = 0, \quad C = C_i \tag{4}$$

Substituting (3) in (2) gives:

$$0 = A \cosh a z_L + B \sinh a z_L$$

and substituting (4) in (2) gives:

$$C_i = A + 0 \text{ and } A = C_i \text{ and } B = -C_i \cosh a z_L / \sinh a z_L$$

$$\therefore \quad C = C_i \cosh az - C_i \frac{\cosh a z_L}{\sinh a z_L} \sinh az$$

$$= \frac{C_i}{\sinh a z_L} (\cosh az \sinh a z_L - \cosh a z_L \sinh az)$$

$$= \frac{C_i \sinh (a z_L - az)}{\sinh a z_L}$$

$$= C_i \frac{\sinh a(z_L - z)}{\sinh a z_L} = C_i \frac{\sinh \sqrt{\alpha/D}(z_L - z)}{\sinh \sqrt{\alpha/D} z_L}$$

Rate of absorption $\quad N_A = -D \left(\dfrac{dC}{dz} \right)_{z=0}$

$$= -D \frac{d}{dz} \left(\frac{\sinh a(z_L - z)}{\sinh a z_L} \right)$$

$$= \frac{D C_i a \cosh a z_L}{\sinh a z_L}$$

$$= D C_i a / \tanh a z_L$$

$$= D C_i a z_L / z_L \tanh a z_L$$

Now $\quad k_L = D/z_L$

$$\beta = \sqrt{D\alpha}/k_L = \sqrt{\alpha/D} \, z_L = a z_L$$

$$\therefore \quad N_A = \frac{k_L C_i \beta}{\tanh \beta}$$

Problem 8.14

The diffusivity of the vapour of a volatile liquid in air can be conveniently determined by Winkelmann's method, in which liquid is contained in a narrow diameter vertical tube maintained at a constant temperature, and an air stream is passed over the top of the tube sufficiently rapidly to ensure the partial pressure of the vapour there remains approximately zero. On the assumption that the vapour is transferred from the surface of the liquid to the air stream by molecular diffusion, calculate the diffusivity of carbon tetrachloride vapour in air at 321 K and atmospheric pressure from the following experimentally obtained data:

Time from commencement of experiment (ks)	Liquid level (cm)
0	0·00
1·6	0·25
11·1	1·29
27·4	2·32
80·2	4·39
117·5	5·47
168·6	6·70
199·7	7·38
289·3	9·03
383·1	10·48

The vapour pressure of carbon tetrachloride at 321 K is $37·6 \, kN/m^2$, and the density of the liquid is $1540 \, kg/m^3$. Take the kilogram molecular volume as $22·4 \, m^3$.

Solution

Equation 8.28 states:

$$N_A = -D \frac{(C_{A2} - C_{A1})}{y_2 - y_1} \frac{C_T}{C_{BM}}$$

In this problem, the distance through which the gas is diffusing will be designated h and $C_{A2} = 0$.

$\therefore$
$$N_A = D(C_A/h)(C_T/C_{BM}) \, kmol/m^2 \, s$$

where C_A is the concentration at the interface.

If the liquid level falls by a distance dh in time dt, the rate of evaporation is given by:

$$N_A = (\rho_L/M) \, dh/dt \, kmol/m^2 \, s$$

Hence
$$(\rho_L/M) \, dh/dt = D(C_A/h)(C_T/C_{BM})$$

If this equation is integrated, noting that when $t = 0$, $h = h_o$, then:

$$h^2 - h_o^2 = (2MD/\rho_L)(C_A C_T/C_{BM})t$$

or, rearranging:

$$t/(h - h_o) = (\rho_L/2MD)(C_{BM}/C_A C_T)(h - h_o) + (\rho_L C_{BM}/MDC_A C_T)h_o$$

Thus a plot of $t/(h - h_o)$ against $(h - h_o)$ will be a straight line of slope s where:

$$s = \rho_L C_{BM}/2MDC_A C_T$$

or
$$D = \rho_L C_{BM}/2MC_A C_T s$$

In this problem the following table may be produced:

t (ks)	1·6	11·1	27·4	80·2	117·5	168·6	199·7	289·3	383·1
$(h - h_0)$ (mm)	2·5	12·9	23·2	43·9	54·7	67·0	73·8	90·3	104·8
$t/(h - h_0)$ (s/m × 10^{-6})	0·64	0·86	1·18	1·83	2·15	3·52	2·71	3·20	3·66

The graph is plotted as Fig. 8b and the slope is found to be:

$$s = (3·54 - 0·5)10^{-6}/100 \times 10^{-3} = 3·04 \times 10^{-7} \text{ s/m}^2$$
$$C_T = (1/22·4)(273/321) = 0·0380 \text{ kmol/m}^3$$
$$M = 154 \text{ kg/kmol}$$
$$C_A = (37·6/101·3)(1/22·4)(273/321) = 0·0141 \text{ kmol/m}^3$$
$$\rho_L = 1540 \text{ kg/m}^3$$
$$C_{B1} = 0·0380 \text{ kmol/m}^3$$
$$C_{B2} = 0·0380 - 0·0141 = 0·0239 \text{ kmol/m}^3$$
$$C_{BM} = (0·0380 - 0·0239)/\ln(0·0380/0·0239)$$
$$= 0·0304 \text{ kmol/m}^3$$

Hence

$$D = 1540 \times 0·0304/(2 \times 154 \times 0·0141 \times 0·0380 \times 3·04 \times 10^{-7})$$
$$= \underline{\underline{9·33 \times 10^{-6} \text{ m}^2/\text{s}}}$$

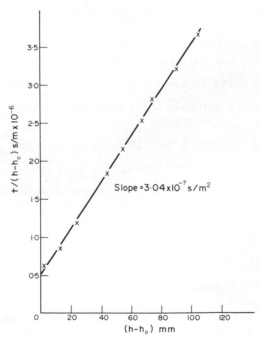

FIG. 8b

Problem 8.15

Ammonia is absorbed in water from a mixture with air using a column operating at atmospheric pressure and 295 K. The resistance to transfer can be regarded as lying entirely within the gas phase. At a point in the column the partial pressure of the ammonia is $6 \cdot 6 \, kN/m^2$. The back pressure at the water interface is negligible and the resistance to transfer can be regarded as lying in a stationary gas film 1 mm thick. If the diffusivity of ammonia in air is $0 \cdot 236 \, cm^2/s$, what is the transfer rate per unit area at that point in the column? If the gas were compressed to $200 \, kN/m^2$ pressure, how would the transfer rate be altered?

Solution

Let subscripts 1 and 2 refer to the water and air side of the stagnant air film respectively and let subscripts A and B refer to ammonia and air.

Equation 8.23 gives the rate of diffusion through a stagnant layer as:

$$N_a = -(D/\mathbf{R}Tx)(P/P_{BM})(P_{A2} - P_{A1})$$

$$P_{A2} = 6 \cdot 6 \, kN/m^2 \quad \text{and} \quad P_{A1} = 0 \, kN/m^2$$

$$P_{B2} = 101 \cdot 3 - 6 \cdot 6 = 94 \cdot 7 \, kN/m^2 \quad \text{and} \quad P_{B1} = 101 \cdot 3 \, kN/m^2$$

$$P_{BM} = 97 \cdot 94 \, kN/m^2 \quad \text{and} \quad P/P_{BM} = 1 \cdot 034$$

Substituting gives
$$N_A = -(0 \cdot 0236 \times 10^{-4}/8 \cdot 314 \times 295 \times 10^{-3})1 \cdot 034(0 - 6 \cdot 6)$$
$$= 6 \cdot 57 \times 10^{-5} \, kmol/m^2 \, s$$

If the pressure is increased to $200 \, kN/m^2$, P_{A1} is still zero, P_{A2} is virtually doubled, P_{B1} and P_{B2} become 200 and $186 \cdot 8 \, kN/m^2$ respectively.

Then
$$P_{BM} = 193 \cdot 3 \, kN/m^2 \quad \text{and} \quad P/P_{BM} = 1 \cdot 034 \text{ as before}$$

From equation 8.44, the diffusivity is inversely proportional to the pressure so that D becomes $0 \cdot 236 \times 10^{-4} \times 101 \cdot 3/200 = 0 \cdot 12 \times 10^{-4} \, m^2/s$.

Hence
$$N_A = -(0 \cdot 12 \times 10^{-4}/8 \cdot 314 \times 295 \times 10^{-3})1 \cdot 017(0 - 13 \cdot 2)$$
$$= 6 \cdot 67 \times 10^{-5} \, kmol/m^2 \, s$$

Thus the rate is substantially unaltered with a virtual doubling of pressure.

Problem 8.16

What are the general principles underlying the two-film penetration and film-penetration theories for mass transfer across a phase boundary? Give the basic differential equations which have to be solved for these theories with the appropriate boundary conditions.

According to the penetration theory, the instantaneous rate of mass transfer per unit area N_A at some time t after the commencement of transfer is given by:

$$N_A = \Delta C \sqrt{\frac{D}{\pi t}}$$

where ΔC is the concentration driving force and D is the diffusivity.

Obtain expressions for the average rates of transfer on the basis of the Higbie and Danckwerts assumptions.

Solution

The various theories for the mechanism of mass transfer across a phase boundary are fully discussed in Section 8.5.

The basic equation for unsteady state equimolecular counter-diffusion is equation 8.50, i.e.:

$$\frac{\partial C_A}{\partial t} = D\left[\left(\frac{\partial^2 C_A}{\partial x^2}\right)_{yz} + \left(\frac{\partial^2 C_A}{\partial y^2}\right)_{xz} + \left(\frac{\partial^2 C_A}{\partial z^2}\right)_{xy}\right]$$

Considering the diffusion of solute A away from the interface in the y-direction the above equation reduces to:

$$\frac{\partial C_A}{\partial t} = D\frac{\partial^2 C_A}{\partial y^2}$$

The boundary conditions which apply are:

$$
\begin{array}{lll}
t = 0 & 0 < y < \infty & C_A = C_o \\
t > 0 & y = 0 & C_A = C_i \\
t > 0 & y = \infty & C_A = C_o
\end{array}
$$

where C_o is the concentration in the bulk of the phase and C_i is the equilibrium concentration at the interface.

The instantaneous rate of mass transfer per unit area N_A at time t is given by:

$$N_A = \Delta C \sqrt{D/\pi t}$$

Higbie assumed that every element of surface is exposed to the gas for the same length of time θ before being replaced by liquid of the bulk composition.

Amount absorbed in time θ,

$$Q = \int_0^\theta N_A \, d\theta$$

$$= \int_0^\theta \Delta C \sqrt{D/\pi\theta} \, d\theta$$

$$= 2\Delta C \sqrt{D\theta/\pi}$$

The average rate of absorption $= Q/\theta$

$$= (2\Delta C\sqrt{D\theta/\pi})/\theta$$
$$= 2\Delta C \sqrt{D/\pi\theta}$$

Danckwerts suggested that each element would not be exposed for the same time but that a random distribution of ages would exist. It is shown in Section 8.5.2 that this age distribution may be expressed $f(t) = s\,e^{-st}$. The average rate of absorption is the value of N_A averaged over all elements of the surface having ages between 0 and ∞ is then given by:

$$\overline{N_A} = s \int_0^\infty N_A\,e^{-s\theta}\,d\theta$$

$$= \Delta C s \sqrt{D/\pi} \int_0^\infty (e^{-s\theta}/\sqrt{\theta})\,d\theta$$

$$= \underline{\Delta C \sqrt{Ds}}$$

Problem 8.17

A solute diffuses from a liquid surface at which its molar concentration is C_i into a liquid with which it reacts. The mass transfer rate is given by Fick's law and the reaction is first order with respect to the solute. In a steady-state process, the diffusion rate falls at a depth L to one half the value at the interface. Obtain an expression for the concentration C of solute at a depth z from the surface in terms of the molecular diffusivity D and the reaction rate constant α. What is the molar flux at the surface?

Solution

As in Problem 8.13 the basic equation is:

$$\frac{d^2C}{dx^2} = a^2 C \tag{1}$$

where $a = \sqrt{\alpha/D}$

Then, $C = A \cosh ax + B \sinh ax$ $\tag{2}$

The first boundary condition is at $x = 0$, $C = C_i$, and $C_i = A$.

Hence $C = C_i \cosh ax + B \sinh ax$ $\tag{3}$

The second boundary condition is that when $x = L$;

$$N_A = -D(\partial C/\partial x)_{x=0} = -2D(\partial C/\partial x)_{x=L}$$

Differentiating (3) gives:

$$dC/dx = C_i a \sinh ax + Ba \cosh ax$$

and $(dC/dx)_{x=0} = Ba$

and $(dC/dx)_{x=L} = aB/2 = C_i a \sinh aL + Ba \cosh aL$

so that $B = \dfrac{2C_i \sinh aL}{1 - 2\cosh aL}$ $\tag{4}$

Substituting (4) into (3) gives:

$$C = C_i \cosh ax + \frac{2C_i \sinh aL \sinh ax}{1 - 2 \cosh aL}$$

$$= C_i[\cosh ax - 2(\cosh ax \cosh aL + \sinh aL \sinh ax)]/(1 - 2 \cosh aL)$$

$$= C_i[\cosh ax - 2 \cosh a(x + L)]$$

The molar flux at the surface $= N_A = -D(dC/dx)_{x=0}$.

$$\frac{dC}{dx} = C_i[a \sinh ax - 2a \sinh a(x + L)]$$

$$(dC/dx)_{x=0} = -2C_i a^2 \sinh aL$$

$$N_A = 2DC_i a^2 \sinh aL$$

$$a = \sqrt{\alpha/D}$$

$$N_A = 2DC_i(\alpha/D) \sinh L\sqrt{\alpha/D}$$

$$= 2C_i\alpha \sinh (L\sqrt{\alpha/D})$$

Problem 8.18

4 cm³ of mixture formed by adding 2 cm³ of acetone to 2 cm³ of dibutyl phthalate is contained in a 6 mm diameter vertical glass tube immersed in a thermostat maintained at 315 K. A stream of air at 315 K and atmospheric pressure is passed over the open top of the tube to maintain a zero partial pressure of acetone vapour at that point. The liquid level is initially 1·15 cm below the top of the tube and the acetone vapour is transferred to the air stream by molecular diffusion alone. The dibutyl phthalate can be regarded as completely non-volatile and the partial pressure of acetone vapour may be calculated from Raoult's law on the assumption that the density of dibutyl phthalate is sufficiently greater than that of acetone for the liquid to be completely mixed.

Calculate the time taken for the liquid level to fall to 5 cm below the top of the tube, neglecting the effects of bulk flow in the vapour.

Kilogram molecular volume = 22·4 m³.

Molecular weights of acetone, dibutyl phthalate = 58 and 279 kg/kmol respectively.

Liquid densities of acetone, dibutyl phthalate = 764 and 1048 kg/m³ respectively.

Vapour pressure of acetone at 315 K = 60·5 kN/m².

Diffusivity of acetone vapour in air at 315 K = 0·123 cm²/s.

Solution

Consider the situation when the liquid has fallen to a depth h cm below the top of the tube.

The volume of acetone evaporated $= (\pi/4)(0\cdot6)^2(h - 1\cdot15)$
$$= 0\cdot283(h - 1\cdot15) \text{ cm}^3$$
At this time, the number of gmol of dibutyl phthalate is given by:

$$2 \times 1\cdot048/278 = 0\cdot00754 \text{ gmol}$$

and the number of gmol acetone by:

$$[2 - 0\cdot283(h - 1\cdot15)]0\cdot764/58 = 0\cdot0306 - 0\cdot00372h$$

$\therefore$ Mol fraction of acetone $= \dfrac{0\cdot0306 - 0\cdot00372h}{0\cdot00754 + 0\cdot0306 - 0\cdot00372h}$

$$= \frac{8\cdot23 - h}{10\cdot24 - h}$$

Partial pressure of acetone $= 60\cdot5 \left(\dfrac{8\cdot23 - h}{10\cdot24 - h}\right) \text{kN/m}^2$

Molar concentration of acetone vapour at the liquid surface

$$= \frac{60\cdot5}{101\cdot3} \times \frac{273}{315} \times \frac{1}{22400} \left(\frac{8\cdot23 - h}{10\cdot24 - h}\right)$$

$$= 2\cdot31 \times 10^{-5} \left(\frac{8\cdot23 - h}{10\cdot24 - h}\right) \text{gmol/cm}^3$$

Rate of evaporation of acetone, $N_A = \dfrac{\mathrm{d}h}{\mathrm{d}t} \times \dfrac{0\cdot764}{58}$

$$= 0\cdot0132(\mathrm{d}h/\mathrm{d}t) \text{ gmol/cm}^2 \text{ s}$$

$$= (D/h) \times \text{molar conc. at surface}$$

$$= (0\cdot123/h)2\cdot31 \times 10^{-5} \left(\frac{8\cdot23 - h}{10\cdot24 - h}\right)$$

$\therefore$ $0\cdot0132 \dfrac{\mathrm{d}h}{\mathrm{d}t} = \dfrac{1}{h} \times 2\cdot84 \times 10^{-6} \left(\dfrac{8\cdot23 - h}{10\cdot24 - h}\right)$

$\therefore$ $\left(\dfrac{10\cdot24 - h}{8\cdot23 - h}\right) h \, \mathrm{d}h = \dfrac{\mathrm{d}t}{4650}$

The time for the liquid level to fall from $1\cdot15$ cm to 5 cm below the top of the tube is found by integrating this equation as:

$$\int_{1\cdot15}^{5} \left(\frac{10\cdot24 - h}{8\cdot23 - h}\right) h \, \mathrm{d}h = \frac{1}{4650} \int_0^t \mathrm{d}t = \int_{1\cdot15}^{5} \left(h - 2\cdot02 - \frac{16\cdot6}{h - 8\cdot23}\right) \mathrm{d}h = \frac{1}{4650} \int_0^t \mathrm{d}t$$

from which $t = 79500 \text{ s}$

$$\equiv \underline{79\cdot5 \text{ ks}} \quad (\approx 22 \text{ hr})$$

THE BOUNDARY LAYER

Problem 9.1

Calculate the thickness of the boundary layer at a distance of 75 mm from the leading edge of a plane surface over which water is flowing at a rate of 3 m/s. Assume that the flow in the boundary layer is streamline and that the velocity u of the fluid at a distance y from the surface can be represented by the relation $u = a + by + cy^2 + dy^3$ (where the coefficients a, b, c, and d are independent of y). Take the viscosity of water as 1 mN s/m².

Solution

At y from the surface, $u = a + by + cy^2 + dy^3$.
When $y = 0$, $u = 0$, and hence $a = 0$.
The shear stress within the fluid, $R_0 = -\mu(\partial u/\partial y)_{y=0}$.
Since $(\partial u/\partial y)$ is constant for small values of y, $(\partial^2 u/\partial y^2)_{y=0} = 0$.
At the edge of the boundary layer, $y = \delta$ and $u = u_s$, the main stream velocity.

$$\partial u/\partial y = 0 \quad \text{and} \quad u = by + cy^2 + dy^3$$
$$\partial u/\partial y = b + 2cy + 3dy^2$$

and
$$\partial^2 u/\partial y^2 = 2c + 6dy$$

When $y = 0$, $\partial^2 u/\partial y^2 = 0$, and hence $c = 0$.
When $y = \delta$, $u = b\delta + d\delta^3 = u_s$

and
$$\partial u/\partial y = b + 3d\delta^2 = 0$$
∴
$$b = -3d\delta^2$$
∴
$$d = -u_s/2\delta^3 \quad \text{and} \quad b = 3u_s/2\delta$$

The velocity profile is given by: $\quad u = (3u_s y/2\delta) - (u_s/2)(y/\delta)^3$

or $\quad u/u_s = 1.5(y/\delta) - 0.5(y/\delta)^3 \quad$ that is equation 9.5.

The integral in the momentum equation 9.2 is now evaluated, and substituting from equations 9.7 and 9.8 into equation 9.2 gives:

$$(\delta/x) = 4.64 Re_x^{-0.5}$$
$$Re_x = (0.075 \times 3 \times 1000/1 \times 10^{-3}) = 225{,}000$$
∴
$$\delta/x = (4.64 \times 225{,}000^{-0.5}) = 0.00978$$

and $\qquad \delta = (0 \cdot 00978 \times 0 \cdot 075) = 0 \cdot 000734$ m

or $\qquad\qquad\qquad \underline{\underline{0 \cdot 734 \text{ mm}}}$

Problem 9.2

Water flows at a velocity of 1 m/s over a plane surface 0·6 m wide and 1 m long. Calculate the total drag force acting on the surface if the transition from streamline to turbulent flow in the boundary layer occurs when the Reynolds group, $Re_x = 10^5$.

Solution

At the far end of the surface, $Re_x = (1 \times 1 \times 1000/1 \times 10^{-3}) = 10^6$.
The mean value of $(R/\rho u_s^2)$ is given by equation 9.34:

$$= 0 \cdot 037(10^6)^{-0 \cdot 2} + (10^6)^{-1}(0 \cdot 646(10^5)^{0 \cdot 5} - 0 \cdot 037(10^5)^{0 \cdot 8})$$
$$= 0 \cdot 00217$$

The total drag force $= (R/\rho u_s^2)(\rho u_s^2)$ (area of the surface)

$$= 0 \cdot 00217 \times 1000 \times 1^2(0 \cdot 6 \times 1)$$
$$= \underline{\underline{1 \cdot 301 \text{ N}}}$$

Problem 9.3

Calculate the thickness of the boundary layer at a distance of 150 mm from the leading edge of a surface over which oil, of viscosity 50 mN s/m² and density 990 kg/m³, flows with a velocity of 0·3 m/s. What is the displacement thickness of the boundary layer?

Solution

$$Re_x = (0 \cdot 150 \times 0 \cdot 3 \times 990)/(50 \times 10^{-3}) = 891$$

For streamline flow in the boundary layer, in equation 9.11:

$$\delta/x = 4 \cdot 64 Re_x^{-0 \cdot 5} = 4 \cdot 64 \times 891^{-0 \cdot 5} = 0 \cdot 155$$

and $\qquad \delta = (0 \cdot 155 \times 0 \cdot 150) = 0 \cdot 0233$ m $\quad$ or $\quad \underline{23 \cdot 3 \text{ mm}}$

In equation 9.13, the displacement thickness, $\delta^* = 0 \cdot 375 \delta$

or $\qquad\qquad \delta^* = (0 \cdot 375 \times 23 \cdot 3) = \underline{\underline{8 \cdot 75 \text{ mm}}}$

Problem 9.4

Calculate the thickness of the laminar sub-layer when benzene flows through a pipe 50 mm diameter at 3000 cm³/s. What is the velocity of the benzene at the edge of the laminar sub-layer? Assume fully developed flow exists within the pipe.

Solution

For benzene, $\rho = 870$ kg/m³, $\mu = 0.7$ mN s/m².
Cross-sectional area of a 50 mm pipe $= (\pi/4) \times 0.050^2 = 0.00196$ m².
Volume flow of benzene $= 3000$ cm³/s or 0.003 m³/s.

$\therefore$ velocity $= (0.003/0.00196) = 1.53$ m/s

and the Reynolds number $Re = (0.050 \times 1.53 \times 870)/(0.7 \times 10^{-3}) = 95,080$
From equation 9.42,

$$\delta_b/d = 62Re^{-0.875} = (62 \times 95,080^{-0.875}) = 2.733 \times 10^{-3}$$

and $\delta_b = (2.733 \times 10^{-3} \times 0.050) = 1.37 \times 10^{-4}$ m   or   0.137 mm

From equation 9.40:

$$u_b/u = 2.49Re^{-0.125} = (2.49 \times 95,080^{-0.125}) = 0.594$$

and $u_b = (0.594 \times 1.53) = 0.91$ m/s

Problem 9.5

Calculate the rise in temperature of water passed at 4 m/s through a smooth 25 mm diameter pipe 6 m long. The water enters at 300 K and the temperature of the wall of the tube can be taken as approximately constant at 330 K. Use:
(a) The simple Reynolds analogy,
(b) The Taylor–Prandtl modification,
(c) The buffer layer equation,
(d) $Nu = 0.023Re^{0.8}Pr^{0.33}$.
Comment on the differences in the results so obtained.

Solution

An approximate solution will be obtained taking the fluid properties at 310 K and neglecting variation with temperature and also assuming that fully developed flow exists.
At 310 K; $\mu = 0.7$ mN s/m²   and $\rho = 1000$ kg/m³.

$\therefore$ $Re = (0.025 \times 4.0 \times 1000)/(0.7 \times 10^{-3}) = 1.429 \times 10^{5}$
$C_p = 4180$ J/kg K   and $k = 0.6$ W/m K.

$\therefore$ $Pr = (4180 \times 0.7 \times 10^{-3}/0.6) = 4.88$

(a) *Reynolds analogy*

In equation 10.70: $\qquad\qquad\qquad\qquad h/C_p\rho u = 0{\cdot}032Re^{-0{\cdot}25}$

$\therefore\qquad\qquad h = (4180 \times 1000 \times 4{\cdot}0)0{\cdot}032(1{\cdot}429 \times 10^5)^{-0{\cdot}25}$

$$= 2{\cdot}75 \times 10^4\,\text{W/m}^2\,\text{K} \quad \text{or} \quad \underline{\underline{27{\cdot}5\,\text{kW/m}^2\,\text{K}}}$$

(b) *Taylor–Prandtl modification*

In equation 10.71: $\qquad h/C_p\rho u = 0{\cdot}032Re^{-0{\cdot}25}/[1 + 2{\cdot}0Re^{-0{\cdot}125}(Pr - 1)]$

$\therefore\qquad\qquad n = 2{\cdot}75 \times 10^4/(1 + 2{\cdot}0 \times 0{\cdot}227 \times 3{\cdot}88)$

$$= 9{\cdot}96 \times 10^3\,\text{W/m}^2\,\text{K} \quad \text{or} \quad \underline{\underline{10{\cdot}0\,\text{kW/m}^2\,\text{K}}}$$

(c) *Buffer layer equation*

In equation 10.72: $\qquad h/C_p\rho u = 0{\cdot}032Re^{-0{\cdot}25}/\{1 + 0{\cdot}82Re^{-0{\cdot}125}$

$$\times [(Pr - 1) + \ln(0{\cdot}833Pr + 0{\cdot}167)]\}$$

$\therefore\qquad\quad h = 2{\cdot}75 \times 10^4/\{1 + 0{\cdot}82 \times 0{\cdot}227[3{\cdot}88 + \ln(3{\cdot}23 + 0{\cdot}167)]\}$

$$= 1{\cdot}41 \times 10^4\,\text{W/m}^2\,\text{K} \quad \text{or} \quad \underline{\underline{14{\cdot}1\,\text{kW/m}^2\,\text{K}}}$$

(d) *Equation 7.50*

$$h \times 0{\cdot}025/0{\cdot}6 = 0{\cdot}023(1{\cdot}429 \times 10^5)^{0{\cdot}8}(4{\cdot}88)^{0{\cdot}33}$$

$\therefore\qquad\qquad h = 0{\cdot}552 \times 1{\cdot}331 \times 10^4 \times 1{\cdot}687$

$$= 1{\cdot}24 \times 10^4\,\text{W/m}^2\,\text{K} \quad \text{or} \quad \underline{\underline{12{\cdot}4\,\text{kW/m}^2\,\text{K}}}$$

Calculation of rise in temperature

In length of pipe dL the rate of heat transfer $= h\pi = 0{\cdot}025\,dL(330 - \theta)$ kW where θ K is the temperature at L m from the inlet.

The rate of increase in the heat content of the water $= 4(\pi/4)0{\cdot}025^2 \times 1000 \times 4{\cdot}18\,d\theta$ kW

and hence the outlet temperature θ' is given by:

$$\int_{300}^{\theta'} d\theta/(330 - \theta) = 0{\cdot}0096h\int_{0}^{6} dL$$

$\therefore\qquad\qquad \ln(330 - \theta') = (\ln 30 - 0{\cdot}0576h) = (3{\cdot}40 - 0{\cdot}0576h)$

The outlet temperature is then calculated in the following table.

	h (kW)	$(3\cdot40 - 0\cdot0576h)$	$(330 - \theta')$	θ' (K)
(a)	27·5	1·816	6·14	323·9
(b)	10·0	2·824	16·84	313·2
(c)	14·1	2·588	13·30	316·7
(d)	12·5	2·680	14·59	315·4

These results confirm that the simple Reynolds analogy is far from accurate when used to calculate the heat transfer to a liquid.

Problem 9.6

Calculate the rise in temperature of a stream of air, entering at 290 K and passing at 4 m/s through the tube maintained at 350 K; other conditions remaining the same as in the previous problem.

Solution

Assuming that the flow is fully developed and taking the physical properties of air at 310 K:

$$\mu = 1\cdot7 \times 10^{-5}\,\text{N s/m}^2, \quad \rho = 1\cdot0\,\text{kg/m}^3, \quad C_p = 1000\,\text{J/kg K}, \quad k = 0\cdot024\,\text{W/m K}$$
$$\therefore \qquad Re = (0\cdot025 \times 4\cdot0 \times 1\cdot0/1\cdot7 \times 10^{-5}) = 5882$$
and $$Pr = (1000 \times 1\cdot7 \times 10^{-5}/0\cdot024) = 0\cdot71$$

Working as in Problem 9.5:

(a) *Reynolds analogy*

$$h = (1000 \times 1\cdot0 \times 4\cdot0)0\cdot032(5882)^{-0\cdot25}$$
$$= 14\cdot6\,\text{W/m}^2\,\text{K} \quad \text{or} \quad \underline{0\cdot0146\,\text{kW/m}^2\,\text{K}}$$

(b) *Taylor–Prandtl equation*

$$h = 14\cdot6/(1 - 2\cdot0 \times 0\cdot339 \times 0\cdot29)$$
$$= 18\cdot2\,\text{W/m}^2\,\text{K} \quad \text{or} \quad \underline{\underline{0\cdot0182\,\text{kW/m}^2\,\text{K}}}$$

(c) *Buffer layer equation*

$$h = 14\cdot6/\{1 + 0\cdot82 \times 0\cdot339[-0\cdot29 + \ln(0\cdot591 + 0\cdot167)]\}$$
$$= 17\cdot3\,\text{W/m}^2\,\text{K} \quad \text{or} \quad \underline{\underline{0\cdot0173\,\text{kW/m}^2\,\text{K}}}$$

(d) *Equation 7.50*

$$h \times 0.025/0.024 = 0.023(5882)^{0.8}(0.71)^{0.33}$$

$$\therefore \qquad h = 0.0221 \times 1037 \times 0.893$$

$$= 20.5 \text{ W/m}^2 \text{ K} \quad \text{or} \quad \underline{0.0205 \text{ kW/m}^2 \text{ K}}$$

Calculation of rise in temperature

In this case:
$$h\pi \times 0.025 \, dL(350 - \theta) = 4(\pi/4)0.025^2 \times 1.0 \times 1.0 \, d\theta$$

and
$$\int_{290}^{\theta'} d\theta/(350 - \theta) = 40h \int_{0}^{6} dL$$

$$\therefore \qquad \ln(350 - \theta') = [\ln(350 - 290) - 240h] = (4.09 - 240h)$$

	h (kW)	$(4.09 - 240h)$	$(350 - \theta')$	θ' (K)
(a)	0.0146	0.586	1.797	348.2
(b)	0.0182	−0.278	0.757	349.2
(c)	0.0173	−0.062	0.940	349.1
(d)	0.0205	−0.830	0.436	349.5

and the Reynolds analogy gives a more acceptable result.

Problem 9.7

Air is flowing at a velocity of 5 m/s over a plane surface. Derive an expression for the thickness of the laminar sub-layer and calculate its value at a distance of 1 m from the leading edge of the surface.

Assume that within the boundary layer outside the laminar sub-layer, the velocity of flow is proportional to the one-seventh power of the distance from the surface and that the shear stress R at the surface is given by:

$$(R/\rho u_s^2) - 0.0228(u_s\rho x/\mu)^{-0.25}$$

where ρ is the density of the fluid (1.3 kg/m³ for air), μ is the viscosity of the fluid (17×10^{-6} N s/m² for air), u_s is the stream velocity (m/s), and x is the distance from the leading edge (m).

Solution

The shear stress in the fluid at the surface, $\qquad R = -\mu u_x/y$

Also from the equation given: $\qquad R = 0.0228\rho u_s^2(\mu/u_s\rho x)^{0.25}$

$$u_x = (0.0228\rho u_s^2 y/\mu)(\mu/u_s\rho x)^{0.25}$$

If the velocity at the edge of the laminar sub-layer is u_b, $u_x = u_b$ when $y = \delta_b$,

$\therefore \qquad u_b = (0 \cdot 0228 \rho u_s^2 \delta_b / \mu)(\mu / u_s \delta \rho)^{0 \cdot 25}$

and $\qquad (\delta_b / \delta) = 43 \cdot 9(u_b / u_s)(\mu / u_s \delta \rho)^{0 \cdot 75}$

The velocity distribution is given by:

$$(\delta_b / \delta)^{1/7} = (u_b / u_s)$$

and hence $\qquad (u_b / u_s)^7 = 43 \cdot 9(u_b / u_s)(\mu / u_s \delta \rho)^{0 \cdot 75}$

or $\qquad (u_b / u_s) = 1 \cdot 87(\mu / u_s \delta \rho)^{0 \cdot 125} = 1 \cdot 87 Re_\delta^{-0 \cdot 125}$

Substituting $0 \cdot 376 x^{0 \cdot 8}(\mu / u_s \rho)^{0 \cdot 2}$ for δ from equation 9.22:

$$(u_b / u_s) = 1 \cdot 87[0 \cdot 376 u_s \rho x^{0 \cdot 8} \mu^{0 \cdot 2} / (\mu u_s^{0 \cdot 2} \rho^{0 \cdot 2})]^{-0 \cdot 125}$$
$$= (1 \cdot 87 / 0 \cdot 376^{0 \cdot 125})(u_s^{0 \cdot 8} x^{0 \cdot 8} \rho^{0 \cdot 8} / \mu^{0 \cdot 8})^{-0 \cdot 125}$$
$$2 \cdot 11 Re_x^{-0 \cdot 1}$$

Now $\qquad (\delta_b / \delta) = (u_b / u_s)^7 = 190 Re_x^{-0 \cdot 7}$

From equation 9.24: $\qquad (\delta / x) = 0 \cdot 376 Re_x^{-0 \cdot 2}$

$\therefore \qquad (\delta_b / x) = (190 \times 0 \cdot 376)/(Re_x^{0 \cdot 7} Re_x^{0 \cdot 2})$
$$= 71 \cdot 5 Re_x^{-0 \cdot 9}$$

In this case, $\qquad Re_x = (1 \times 5 \times 1 \cdot 3/17 \times 10^{-6}) = 3 \cdot 82 \times 10^5$
$$\delta_b = 1 \cdot 0 \times 71 \cdot 5(3 \cdot 82 \times 10^5)^{-0 \cdot 9}$$
$$= 6 \cdot 77 \times 10^{-4} \, \text{m} \quad \text{or} \quad \underline{\underline{0 \cdot 677 \, \text{mm}}}$$

Problem 9.8

Air flows through a smooth circular duct of internal diameter $0 \cdot 25$ m at an average velocity of 15 m/s. Calculate the fluid velocity at points 50 mm and 5 mm from the wall. What will be the thickness of the laminar sub-layer if this extends to $u^+ = y^+ = 5$? The density of air may be taken as $1 \cdot 12$ kg/m^3 and its viscosity as $0 \cdot 02$ mN s/m^2.

Solution

Reynolds number, $\qquad Re = (0 \cdot 25 \times 15 \times 1 \cdot 12/0 \cdot 02 \times 10^{-3}) = 2 \cdot 1 \times 10^5$

and hence, from figure 3.7: $\qquad (R/\rho u^2) = 0 \cdot 0019$

$$u = 0 \cdot 82 u_s$$

$\therefore \qquad u_s = (15/0 \cdot 82) = 18 \cdot 3 \, \text{m/s}$

$$u^* = u \sqrt{(R/\rho u^2)} = 15 \sqrt{(0 \cdot 0019)} = 0 \cdot 65 \, \text{m/s}$$

At 50 mm from the wall

$$(y/r) = (0\cdot050/0\cdot125) = 0\cdot4$$

and from equation 9.49: $u_x = u_s + 2\cdot5u^* \ln (y/r)$

$$= 18\cdot3 + 2\cdot5 \times 0\cdot65 \ln 0\cdot4 = \underline{\underline{16\cdot8 \text{ m/s}}}$$

At 5 mm from the wall

$$(y/r) = (0\cdot005/0\cdot125) = 0\cdot04$$

and $u_x = 18\cdot3 + 2\cdot5 \times 0\cdot65 \ln 0\cdot04 = \underline{\underline{13\cdot1 \text{ m/s}}}$

From equation 9.59, the thickness of the lamina sub-layer is given by:

$$\delta_b = 5dRe^{-1}(R/\rho u^2)^{-0\cdot5}$$
$$= 5 \times 0\cdot25(2\cdot1 \times 10^5)^{-1}(0\cdot0019)^{-0\cdot5}$$
$$= 1\cdot37 \times 10^{-4} \text{ m} \quad \text{or} \quad \underline{\underline{0\cdot0137 \text{ mm}}}$$

Problem 9.9

Obtain the momentum equation for an element of the boundary layer. If the velocity profile in the laminar region can be represented approximately by a sine function, calculate the boundary layer thickness in terms of distance from the leading edge of the surface.

Solution

The deviation of the momentum equation for an element of the boundary layer is presented in detail in section 9.2. The final expression is:

$$-R_0 = \rho\partial \left(\int_0^l (u_s - u_x)u_x \, dy \right) \Big/ \partial x$$

A sine function is developed as follows:

when $y = 0, \quad u_x = 0$
when $y = \delta, \quad u_x = u_s$

Thus $u_x = u_s \sin (ay)$

$\therefore$ when $y = \delta, \quad \sin ay = \pi/2 \quad$ or $\quad a\delta = \pi/2 \quad$ and $\quad a = \pi/2\delta$.
The function is therefore:

$$u_x = u_s \sin (\pi y/2d)$$

and over the range $0 < y < \delta$

$$(u_x/u_s) = \sin [\pi/2(y/\delta)]$$

The integral in the momentum equation may now be evaluated for the laminar boundary layer considering the ranges $0 < y < \delta$ and $\delta < y < l$ separately.

$$\therefore \quad \int_0^l (u_s - u_x)u_x \, dy = \int_0^\delta u_s^2\{1 - \sin[\pi/2(y/\delta)]\}\{\sin[\pi/2(y/\delta)]\} \, dy + \int_\delta^l (u_s - u_s)u_s \, dy$$

$$= u_s^2 \int_0^\delta [\sin(\pi y/2\delta) - \sin^2(\pi y/2\delta)] \, dy$$

$$= u_s^2[-\cos(\pi y/2\delta)/(\pi/2\delta) + y/2 - \sin(\pi^y/\delta)/(2\pi/\delta)]_0^\delta$$

$$= u_s^2\delta[(2/\pi) - (\tfrac{1}{2})]$$

Now $\qquad\qquad R_0 = -\mu(\partial u_x/\partial y)_{y=0} = -\mu u_s \pi/2\delta$

Substituting in the momentum equation:

$$\rho\partial[u_s^2\delta(2/\pi - \tfrac{1}{2})]/\partial x = \mu u_s \pi/2\delta$$

$$\therefore \qquad\qquad \delta\,d\delta = \mu\pi^2 \, dx/\rho u_s(4 - \pi)$$

$$\therefore \qquad\qquad \delta^2/z = [\pi^2/(4 - \pi)](\mu x/\rho u_s)$$

and $\qquad\qquad \delta = 4{\cdot}80(\mu x/\rho u_s)^{0\cdot5}$

$$\therefore \qquad\qquad (\delta/x) = 4{\cdot}80(\mu/x\rho u_s)^{0\cdot5} = \underline{\underline{4{\cdot}80Re_x^{-0\cdot5}}}$$

HEAT, MASS AND MOMENTUM TRANSFER

Problem 10.1

If the temperature rise per metre length along a pipe carrying air at 12·2 m/s is 66 K, what will be the corresponding pressure drop for a pipe temperature of 420 K and an air temperature of 310 K?

The density of air at 310 K is 1·14 kg/m³.

Solution

For a pipe diameter d m, the mass flow $= u\rho\pi d^2/4$

and the rate of heat transfer, $q = (u\rho\pi d^2/4)C_p\Delta t$ W/m where Δt is the temperature rise (K/m).

Also $q = hA(t_w - t_m) = h\pi dL(t_w - t_m)$ W where t_w and t_m are the mean wall and fluid temperatures.

Thus, for $L = 1$ m:

$$h/C_p u = d\Delta t/4(t_w - t_m)$$

Therefore from equation 10.38:

$$R/\rho u^2 = d\Delta t/4(t_w - t_m)$$

Substituting in equation 3.15:

$$\Delta P = 4d\Delta t(L/d)\rho u^2/4(t_w - t_m)$$
$$= \Delta t\rho u^2 L/(t_w - t_m)$$
$$= (66 \times 1·14 \times 12·2^2 \times 1·0)/(420 - 310)$$
$$= \underline{\underline{101·8 \ (\text{N/m}^2)/\text{m}}}$$

Problem 10.2

It is required to warm a quantity of air from 289 K to 313 K by passing it through a number of parallel metal tubes of inner diameter 50 mm maintained at 373 K. The pressure drop must not exceed 250 N/m². How long should the individual tubes be?

The density of air at 301 K is 1·19 kg/m³ and the coefficients of heat transfer by convection from the tube to air are 45, 62, and 77 W/m² K for velocities of 20, 24, and 30 m/s at 301 K respectively.

Solution

From equations 10.38 and 3.15:

$$\Delta P = 4(h/C_p\rho u)(l/d)\rho u^2 = 4hlu/C_p d$$
$$\therefore \qquad 250 = 4(hlu/C_p \times 0{\cdot}050)$$

or $$h = 3{\cdot}125 C_p/lu \qquad\qquad\qquad (i)$$

The heat transferred to the air $= u(\pi d^2/4)\rho C_p(T_2 - T_1)$
$$= u(\pi \times 0{\cdot}050^2/4)1{\cdot}19 C_p(313 - 289)$$
$$= 0{\cdot}056 C_p u \text{ W}$$

This is equal to $\quad h\pi dl(T_w - T_m) = h\pi \times 0{\cdot}0501(373 - 301) = 11{\cdot}31hl \text{ W}$

$$\therefore \qquad\qquad 11{\cdot}31hl = 0{\cdot}056 C_p u$$

or $$h = 0{\cdot}0050 C_p u/l \qquad\qquad\qquad (ii)$$

From (i), $$\qquad (C_p/l) = 0{\cdot}32hu$$

and in (ii), $$\qquad h = 0{\cdot}005 \times 0{\cdot}32hu^2$$

$$\therefore \qquad\qquad u = 25 \text{ m/s}$$

Interpolation of the data given indicates a value of $h = 64 \text{ W/m}^2 \text{ K}$.

$$\therefore \qquad (C_p/l) = (0{\cdot}32 \times 64 \times 25) = 512 \text{ J/m K kg}$$

For air: $$\qquad\qquad C_p = 1000 \text{ J/kg K}$$

and hence $$\qquad l = (1000/512) = \underline{\underline{1{\cdot}95 \text{ m}}}$$

Problem 10.3

Air at 330 K, flowing at 10 m/s, enters a pipe of inner diameter 25 mm, maintained at 415 K. The drop of static pressure along the pipe is 80 N/m² per metre length. Using the Reynolds analogy between heat transfer and friction, estimate the temperature of the air 0·6 m along the pipe.

Solution

Over a length of 0·6 m: $$\qquad \Delta P = (0{\cdot}6 \times 80) = 48 \text{ N/m}^2$$

From equations 10.38 and 3.15:

$$48 = 4(h/C_p\rho u)(0{\cdot}6/0{\cdot}025)\rho \times 10^2$$

and $$\qquad h/C_p\rho u = 0{\cdot}005/\rho$$

The heat transferred to the air $= u(\pi d^2/4)\rho C_p(T_2 - T_1)$
$$= 10(\pi \times 0{\cdot}025^2/4)\rho C_p(T_2 - 330)$$
$$= 0{\cdot}0049\rho C_p(T_2 - 330)$$

This is also equal to $h\pi dl(T_w - T_m) = h\pi \times 0{\cdot}025 \times 0{\cdot}6(415 - 0{\cdot}5)(T_2 + 330)$
$$= 0{\cdot}047h(250 - 0{\cdot}5 T_2)$$

$\therefore$ $(T_2 - 330)/(250 - 0.5T_2) - 9.59h/C_p\rho - 9.59(0.005u/\rho)$

Taking the density of air as 1.10 kg/m^3:

$$9.59(0.005u/\rho) = (9.59 \times 0.005 \times 10/1.10) = 0.436$$

and hence $T_2 = 360$ K

Problem 10.4

Air flows at 12 m/s through a pipe of inside diameter 25 mm. The rate of heat transfer by convection between the pipe and the air is 60 W/m^2 K. Neglecting the effects of temperature variation, estimate the pressure drop per metre length of pipe.

Solution

From equations 10.38 and 3.15:

$$\Delta P = 4(h/C_p\rho u)(l/d)\rho u^2$$

Taking $C_p = 1000$ J/kg K and $l = 1$ m,

$$\Delta P = 4(60/1000\rho \times 12)(1.0/0.025)\rho \times 12^2$$
$$= 115.2 \text{ N/m}^2$$

Problem 10.5

Apply Reynolds analogy to solve the following problem. Air at 320 K and atmospheric pressure is flowing through a smooth pipe of 50 mm internal diameter and the pressure drop over a 4 m length is found to be 1.5 kN/m^2. By how much would you expect the air temperature to fall over the first metre of pipe length if the wall temperature there is kept constant at 295 K?

Viscosity of air = 0.018 mN s/m^2. Specific heat of air = 1.05 kJ/kg K.

Solution

(In essence, this is the same as Problem 7.40 and an alternative solution is presented here.)

From equations 10.38 and 3.15: $\Delta P = 4(h/C_p\rho u)(l/d)\rho u^2$.

For a length of 4 m: $1500 = 4(h/C_p\rho u)(4.0/0.050)\rho u^2$.

$\therefore$ $hu/C_p = 4.69$ kg/m s^2 (i)

The rate of heat transfer is given by $h\pi dl(T_m - T_w)$, which for a length of 1 m

$$= h\pi \times 0.050 \times 1.0(0.5(320 + T_2) - 295) = 0.157h(0.5T_2 - 135)$$

The heat lost by the air $= u(\pi d^2/4)\rho C_p(T_1 - T_2)$

$$= u(\pi \times 0.050^2/4)\rho C_p(320 - T_2) = 0.00196u\rho C_p(320 - T_2)$$

$\therefore$

$$80.1(h/C_p\rho u) = (320 - T_2)/(0.5T_2 - 135)$$

Substituting from (i):

$$80.1(4.69/\rho u^2) = (320 - T_2)/(0.5T_2 - 135) \tag{ii}$$

From equation 10.70:

$$h/C_p\rho u = 0.032(\mu/du\rho)^{0.25}$$

At 320 K and 101·3 kN/m^2: $\qquad \rho = (28.9/22.4)(273/320) = 1.10 \text{ kg/m}^3$

$\therefore$

$$4.69/(1.10u^2) = 0.032[0.018 \times 10^{-3}/(0.050 \times 1.10u)]^{0.25}$$

$\therefore$

$$u = 51.4 \text{ m/s}$$

Hence, in (ii):

$$(80.1 \times 4.69)/(1.10 \times 51.4^2) = (320 - T_2)/(0.5T_2 - 135)$$

and

$$T_2 = 316 \text{ K}$$

The temperature drop over the first metre is therefore 4 K.

(Note: in these problems, an arithmetic mean temperature difference is used rather than a logarithmic value for ease of solution. This is probably justified in view of the small temperature changes involved and also the approximate nature of the Reynolds analogy.)

Problem 10.6

Obtain an expression for the simple Reynolds analogy between heat transfer and friction. Indicate the assumptions which are made in the derivation and the conditions under which you would expect the relation to be applicable.

The Reynolds number of a gas flowing at 2·5 kg/m^2 s through a smooth pipe is 20 000. If the specific heat of the gas at constant pressure is 1·67 kJ/kg K, what will the heat transfer coefficient be?

Solution

The derivation of the simple Reynolds analogy and its application is presented in detail in Section 10.4.

For a Reynolds number of 2.0×10^4, $(R/\rho u^2) = 0.0032$ for a smooth pipe (Fig. 3.7).

From equation 10.38: $\qquad (h/C_p\rho u) = 0.0032$

$$\rho u = 2.5 \text{ kg/m}^2$$

and hence, $\qquad h = (0.0032 \times 1670 \times 2.5) = 13.4 \text{ W/m}^2 \text{ K}$

SECTION 11

HUMIDIFICATION AND WATER COOLING

Problem 11.1

In a process in which benzene is used as a solvent, it is evaporated into dry nitrogen. The resulting mixture at a temperature of 297 K and a pressure of $101 \cdot 3$ kN/m² has a relative humidity of 60%. It is required to recover 80% of the benzene present by cooling to 283 K and compressing to a suitable pressure. What should this pressure be?

Vapour pressure of benzene: at 297 K = $12 \cdot 2$ kN/m²; at 283 K = $6 \cdot 0$ kN/m².

Solution

From the definition of percentage relative humidity (Section 11.2.1):

$$P_w = P_{w_0} \times \mathrm{RH}/100$$

At 297 K;
$$P_w = (12 \cdot 2 \times 60/100) = 7 \cdot 32 \text{ kN/m}^2$$

In the benzene-nitrogen mixture:

mass of benzene $= P_w M_w / \mathbf{R} T = (7 \cdot 32 \times 78)/(8 \cdot 314 \times 297) = 0 \cdot 231$ kg/m³
mass of nitrogen $= (P - P_w) M_A / \mathbf{R} T = (101 \cdot 3 - 7 \cdot 32)28/(8 \cdot 314 \times 297)$
$\qquad\qquad = 1 \cdot 066$ kg/m³

where the molecular weights of benzene and nitrogen are 78 and 28 kg/kmol respectively.

The humidity is, therefore, $\mathscr{H} = (0 \cdot 231/1 \cdot 066) = 0 \cdot 217$ kg/kg dry nitrogen.

In order to recover 80% of the benzene, the humidity must be reduced by 80%.

As the vapour will be in contact with liquid benzene, the nitrogen will be saturated with benzene vapour and hence, at 283 K:

$$\mathscr{H}_0 = [0 \cdot 217(100 - 80)/100] = 0 \cdot 0433 \text{ kg/kg dry nitrogen}$$

Thus in equation 11.2:

$$0 \cdot 0433 = [6 \cdot 0/(P - 6 \cdot 0)](78/28)$$

and
$$\underline{\underline{P = 392 \text{ kN/m}^2}}$$

Problem 11.2

$0.6 \, \mathrm{m^3/s}$ of gas is to be dried from a dew point of 294 K to a dew point of 277.5 K. How much water must be removed and what will be the volume of the gas after drying?

Vapour pressure of water at $294 \, \mathrm{K} = 2.5 \, \mathrm{kN/m^2}$. Vapour pressure of water at $277.5 \, \mathrm{K} = 0.85 \, \mathrm{kN/m^2}$.

Solution

When the gas is cooled to 294 K, it will be saturated and $P_{w_0} = 2.5 \, \mathrm{kN/m^2}$.

From section 11.2, mass of vapour $= P_{w_0} M_w / \mathbf{R} T$

$$= (2.5 \times 18)/(8.314 \times 294) = 0.0184 \, \mathrm{kg/m^3 \, gas}.$$

When water has been removed, the gas will be saturated at 277.5 K, and $P_w = 0.85 \, \mathrm{kN/m^2}$.

At this stage,   mass of vapour $= (0.85 \times 18)/(8.314 \times 277.5)$

$$= 0.0066 \, \mathrm{kg/m^3 \, gas}$$

Hence,   water to be removed $= (0.0184 - 0.0066) = 0.0118 \, \mathrm{kg/m^3 \, gas}$

or $(0.0118 \times 0.6) = \underline{0.00708 \, \mathrm{kg/s}}$

Assuming the gas flow, $0.6 \, \mathrm{m^3/s}$ is referred to 273 K and $101.3 \, \mathrm{kN/m^2}$, $0.00708 \, \mathrm{kg/s}$ water is equivalent to $(0.00708/18) = 3.933 \times 10^{-4} \, \mathrm{kmol/s}$.

1 kmol of vapour occupies $22.4 \, \mathrm{m^3}$ at STP,

and hence   volume of water removed $= (3.933 \times 10^{-4} \times 22.4) = 8.81 \times 10^{-3} \, \mathrm{m^3/s}$

or $0.00881 \, \mathrm{m^3/s}$

Hence gas flow after drying $= (0.60 - 0.00881) = \underline{0.591 \, \mathrm{m^3/s} \text{ at STP}}$.

Problem 11.3

Wet material, containing 70% moisture, is to be dried at the rate of $0.15 \, \mathrm{kg/s}$ in a counter-current dryer to give a product containing 5% moisture (both on the wet basis). The drying medium consists of air heated to 373 K and containing water vapour equivalent to a partial pressure of $1.0 \, \mathrm{kN/m^2}$. The air leaves the dryer at 313 K and 70% saturated. Calculate how much air will be required to remove the moisture. The vapour pressure of water at 313 K may be taken as $7.4 \, \mathrm{kN/m^2}$.

Solution

The feed is $0.15 \, \mathrm{kg/s}$ wet material containing 0.70 kg water/kg feed.

Thus water in feed $= (0.15 \times 0.70) = 0.105 \, \mathrm{kg/s}$ and dry solids $= (0.15 - 0.105) = 0.045 \, \mathrm{kg/s}$.

The product contains 0.05 kg water/kg product. Thus if w kg/s is the amount of water in the product,

$$w/(w + 0.045) = 0.05 \quad \text{or} \quad w = 0.00237 \text{ kg/s}$$

and water to be removed $= (0.105 - 0.00237) = 0.1026$ kg/s.

The inlet air is at 373 K and the partial pressure of the water vapour is 1 kN/m². Assuming a total pressure of 101·3 kN/m², the humidity from equation 11.1 is:

$$\mathscr{H}_1 = [P_w/(P - P_w)](M_w/M_A) \tag{11.1}$$
$$= [1.0/(101.3 - 1.0)](18/29) = 0.0062 \text{ kg/kg dry air}$$

The outlet air is at 313 K and is 70% saturated. Thus from Problem 11.1:

$$P_w = P_{w_0} \times RH/100$$
$$= (7.4 \times 70/100) = 5.18 \text{ kN/m}^2$$

and $\qquad \mathscr{H}_2 = [5.18/(101.3 - 5.18)](18/29) = 0.0335$ kg/kg dry air

The increase in humidity is $(0.0335 - 0.0062) = 0.0273$ kg/kg dry air and this must correspond to the water removed, 0·1026 kg/s. Thus if m kg/s is the mass flow of dry air,

$$0.0273m = 0.1026 \quad \text{and} \quad m = 3.76 \text{ kg/s dry air}$$

In the inlet air, this is associated with 0·0062 kg water vapour, or

$$(0.0062 \times 3.76) = 0.0233 \text{ kg/s}$$

Hence, the mass of moist air required at the inlet conditions $= (3.76 + 0.0233)$
$$= 3.783 \text{ kg/s}$$

Problem 11.4

30 000 m³ of coal gas (measured at 289 K and 101·3 kN/m² saturated with water vapour) is compressed to 340 kN/m² pressure, cooled to 289 K and the condensed water is drained off. Subsequently the pressure is reduced to 170 kN/m² and the gas is distributed at this pressure and 289 K. What is the percentage humidity after this treatment?

The vapour pressure of water at 289 K is 1·8 kN/m².

Solution

At 289 K and 101·3 kN/m², the gas is saturated and $P_{w_0} = 1.8$ kN/m².

Thus from equation 11.2: $\mathscr{H}_0 = [1.8/(101.3 - 1.8)]18/M_A = 0.3256/M_A$ kg/kg dry gas

At 289 K and 340 kN/m², the gas is in contact with condensed water and therefore still saturated. Thus $P_{w_0} = 1.8$ kN/m² and

$$\mathscr{H}_0 = [1.8/(340 - 1.8)]18/M_A = 0.0958/M_A \text{ kg/kg dry gas}$$

At 289 K and 170 kN/m^2, the humidity is the same, and in equation 11.2:

$$0 \cdot 0958/M_A = [P_w/(170 - P_w)]18/M_A$$

or
$$P_w = 0 \cdot 90 \text{ kN/m}^2$$

The percentage humidity is given by equation 11.3:

$$= [(P - P_{wo})/(P - P_w)](100 P_w/P_{wo})$$
$$= [(170 - 1 \cdot 8)/(170 - 0 \cdot 90)](100 \times 0 \cdot 90/1 \cdot 8)$$
$$= \underline{\underline{49 \cdot 73\%}}$$

Problem 11.5

A rotary counter-current dryer is fed with ammonium nitrate containing 5% moisture at a rate of 1·5 kg/s and discharges the nitrate with 0·2% moisture. The air enters at 405 K and leaves at 355 K; the humidity of the entering air being 0·007 kg moisture/kg dry air. The nitrate enters at 294 K and leaves at 339 K.

Neglecting radiation losses, calculate the weight of dry air passing through the dryer and the humidity of the air leaving the dryer.

Latent heat of water at 294 K = 2450 kJ/kg.
Specific heat of ammonium nitrate = 1·88 kJ/kg K.
Specific heat of dry air = 0·99 kJ/kg K.
Specific heat of water vapour = 2·01 kJ/kg K.

Solution

The feed of wet nitrate is 1·5 kg/s containing 5·0% moisture or $(1 \cdot 5 \times 5/100) = 0 \cdot 075$ kg/s water.

$$\therefore \qquad \text{flow of dry solids} = (1 \cdot 5 - 0 \cdot 075) = 1 \cdot 425 \text{ kg/s}$$

If the product contains w kg/s water, then

$$w/(w + 1 \cdot 425) = 0 \cdot 2/100 \quad \text{or} \quad w = 0 \cdot 00286 \text{ kg/s}$$

and the water evaporated = $(0 \cdot 075 - 0 \cdot 00286) = 0 \cdot 07215$ kg/s.

The problem now consists of an enthalpy balance around the unit, and for this purpose a datum temperature of 294 K will be chosen. It will be assumed that the flow of dry air into the unit is m kg/s.

Considering the inlet streams

(i) Nitrate: this enters at the datum of 294 K and hence the enthalpy = 0.
(ii) Air: m kg/s of dry air is associated with 0·007 kg moisture/kg dry air.

$$\therefore \qquad \text{enthalpy} = [(m \times 0 \cdot 99) + (0 \cdot 007 m \times 2 \cdot 01)](405 - 294) = 111 \cdot 5 \, m \text{ kW}$$

and the total heat into the system = $111 \cdot 5 \, m$ kW.

Considering the outlet streams

(i) Nitrate: 1·425 kg/s dry nitrate contains 0·00286 kg/s water and leaves the unit at 339 K.

$$\therefore \qquad \text{enthalpy} = [(1\cdot425 \times 1\cdot88) + (0\cdot00286 \times 4\cdot18)](339 - 294) = 120\cdot7 \text{ kW}$$

(ii) air: the air leaving contains $0\cdot007m$ kg/s water from the inlet air plus the water evaporated. It will be assumed that evaporation takes place at 294 K.
Thus:
enthalpy of dry air $= m \times 0\cdot99(355 - 294) = 60\cdot4m$ kW
enthalpy of water from inlet air $= 0\cdot007m \times 2\cdot01(355 - 294) = 0\cdot86m$ kW
enthalpy in the evaporated water $= 0\cdot07215[2450 + 2\cdot01(355 - 294)] = 185\cdot6$ kW
the total heat out of the system, neglecting losses $= 306\cdot3 + 61\cdot3m$ kW.

Making a balance:

$$111\cdot5m = 306\cdot3 + 61\cdot3m$$

or $$m = 6\cdot10 \text{ kg/s dry air}$$

Thus, including the moisture in the inlet air, moist air fed to the dryer

$$= 6\cdot10(1 + 0\cdot007) = \underline{\underline{6\cdot15 \text{ kg/s}}}$$

Water entering with the air $= 6\cdot10 \times 0\cdot007 = 0\cdot0427$ kg/s.
Water evaporated $= 0\cdot07215$ kg/s.
Water leaving with the air $= (0\cdot0427 + 0\cdot07215) = 0\cdot1149$ kg/s
Humidity of outlet air $= (0\cdot1149/6\cdot10) = \underline{\underline{0\cdot0188 \text{ kg/kg dry air}}}$.

Problem 11.6

Material is fed to a dryer at the rate of 0·3 kg/s and the moisture removed is 35% of the wet charge. The stock enters and leaves the dryer at 324 K. The air temperature falls from 341 K to 310 K; its humidity rising from 0·01 to 0·02 kg/kg. Calculate the heat loss to the surroundings.
Latent heat of water at 324 K = 2430 kJ/kg.
Specific heat of dry air = 0·99 kJ/kg K.
Specific heat of water vapour = 2·01 kJ/kg K.

Solution

The wet fed is 0·3 kg/s and the water removed is 35%, or:

$$(0\cdot3 \times 35/100) = 0\cdot105 \text{ kg/s}$$

If the flow of dry air is m kg/s, the increase in humidity $= (0\cdot02 - 0\cdot01) = 0\cdot01$ kg/kg

or $$0\cdot01m = 0\cdot105 \quad \text{and} \quad m = 10\cdot5 \text{ kg/s}$$

This completes the mass balance, and the problem is now to make an enthalpy balance along the lines of Problem 11.5. As the stock enters and leaves at 324 K, no heat is transferred from the air and the heat lost by the air must represent the heat used for evaporation plus the heat losses, say L kW.

Thus heat lost by the inlet air and associated moisture

$$= [(10\cdot5 \times 0\cdot99) + (0\cdot01 \times 10\cdot5 \times 2\cdot01)](341 - 310) = 328\cdot8 \text{ kW}$$

Heat leaving in the evaporated water $= 0\cdot105[2430 + 2\cdot01(310 - 324)] = 252\cdot2$ kW.

The difference between the inlet and outlet stock streams is represented by the sensible heat in the water evaporated or

$$0\cdot105 \times 4\cdot18(324 - 310) = 6\cdot1 \text{ kW}$$

Making a balance,

$$328\cdot8 = 252\cdot2 + 6\cdot1 + L \quad \text{or} \quad \underline{\underline{L = 70\cdot5 \text{ kW}}}$$

Problem 11.7

A rotary dryer is fed with sand at a rate of 1 kg/s. The feed is 50% wet and the sand is discharged with 3% moisture. The entering air is at 380 K and has an absolute humidity of 0·007 kg/kg. The wet sand enters at 294 K and leaves at 309 K and the air leaves at 310 K. Calculate the mass of air passing through the dryer and the humidity of the air leaving the dryer. Allow a radiation loss of 25 kJ/kg dry air.

Latent heat of water at 294 K = 2450 kJ/kg
Specific heat of sand   = 0·88 kJ/kg K
Specific heat of dry air   = 0·99 kJ/kg K
Specific heat of vapour   = 2·01 kg K

Solution

The feed of wet sand is 1 kg/s containing 50% moisture or $(1\cdot0 \times 50/100) = 0\cdot50$ kg/s water.

∴   flow of dry sand $= (1\cdot0 - 0\cdot5) = 0\cdot50$ kg/s

If the dried sand contains w kg/s water, then

$$w/(w + 0\cdot50) = 3\cdot0/100 \quad \text{or} \quad w = 0\cdot0155 \text{ kg/s}$$

and the water evaporated $= (0\cdot50 - 0\cdot0155) = 0\cdot4845$ kg/s.

Assuming a flow of m kg/s dry air, then a heat balance may be made based on a datum temperature of 294 K.

Considering the inlet streams

(i) Sand: this enters at 294 K and hence the enthalpy $= 0$.
(ii) Air: m kg/s of dry air is associated with 0·007 kg/kg moisture.

∴   enthalpy $= [(m \times 0\cdot99) + (0\cdot007m \times 2\cdot01)](380 - 294) = 86\cdot4m$ kW

and the total heat into the system $= 86\cdot4m$ kW.

Considering the outlet streams

(i) Sand: 0·50 kg/s dry sand contains 0·0155 kg/s water and leaves the unit at 309 K.

$$\therefore \qquad \text{enthalpy} = [(0\cdot5 \times 0\cdot88) + (0\cdot0155 \times 4\cdot18)](309 - 294) = 7\cdot6 \text{ kW}$$

(ii) Air: the air leaving contains $0\cdot07m$ kg/s water from the inlet air plus the water evaporated. It will be assumed that evaporation takes place at 294 K. Thus:

enthalpy of dry air $= m \times 0\cdot99(310 - 294) = 15\cdot8m$ kW

enthalpy of water from inlet air $= 0\cdot007m \times 2\cdot01(310 - 294) = 0\cdot23m$ kW

enthalpy in the evaporated water $= 0\cdot4845[2430 + 2\cdot01(310 - 294)] = 1192\cdot9$ kW

a total of $(16\cdot03m + 1192\cdot9)$ kW

(iii) Radiation losses: $\qquad\qquad$ loss $= 25$ kJ/kg dry air $\quad$ or $\quad 25m$ kW

and the total heat out $= 41\cdot03m + 1200\cdot5$ kW.

Making a balance:

$$86\cdot4m = 41\cdot03m + 1200\cdot5 \quad \text{or} \quad m = 26\cdot5 \text{ kg/s}$$

Thus the flow of dry air through the dryer $= \underline{\underline{26\cdot5 \text{ kg/s}}}$

and the flow of inlet air $= (26\cdot5 \times 1\cdot007) = \underline{\underline{26\cdot7 \text{ kg/s}}}$

As in Problem 11.5, water leaving with the air $= (26\cdot5 \times 0\cdot007) + 0\cdot4845 = 0\cdot67$ kg/s

and humidity of the outlet air $= (0\cdot67/26\cdot5) = \underline{\underline{0\cdot025 \text{ kg/kg}}}$.

Problem 11.8

Water is to be cooled in a packed tower from 330 to 295 K by means of air flowing counter currently. The liquid flows at a rate of 275 cm³/m² s and the air at 0·7 m³/m² s. The entering air has a temperature of 295 K and a relative humidity of 20%. Calculate the required height of tower and the condition of the air leaving at the top.

The whole of the resistance to heat and mass transfer can be considered as being within the gas phase and the product of the mass transfer coefficient and the transfer surface per unit volume of column $(h_D a)$ can be taken as $0\cdot2 \text{ s}^{-1}$.

Solution

Assuming, the latent heat of water at 273 K $= 2495$ kJ/kg

$\qquad\qquad$ specific heat of dry air $\qquad = 1\cdot003$ kJ/kg K

$\qquad\qquad$ specific heat of water vapour $\quad = 2\cdot006$ kJ/kg K

the enthalpy of the inlet air stream

$$H_{G_1} = 1\cdot003(295 - 273) + \mathcal{H}(2495 + 2\cdot006(295 - 273))$$

From Figure 11.4, when $\theta = 295$ K and 20% RH, $\mathcal{H} = 0.003$ kg/kg, and

$$H_{G_1} = (1.003 \times 22) + 0.003(2495 + (2.006 \times 22)) = 29.68 \text{ kJ/kg}$$

In the inlet air, the humidity is 0.003 kg/kg dry air or $(0.003/18)/(1/29) = 0.005$ kmol/kmol dry air.

Hence the flow of dry air $= (1 - 0.005)0.70 = 0.697$ m^3/m^2 s.

Density of air at 295 K $= (29/22.4)(273/295) = 1.198$ kg/m^3.

and hence the mass flow of dry air $= (0.697 \times 1.198) = 0.835$ kg/m^2 s

and the mass flow of water $= 275 \times 10^{-6}$ m^3/m^2 s or $275 \times 10^{-6} \times 1000 = 0.275$ kg/m^2 s.

The slope of the operating line, given by Equation 11.37 is

$$LC_L/G = 0.275 \times 4.18/0.835 = 1.38$$

The coordinates of the bottom of the operating line are:

$$\theta_{L1} = 295 \text{ K} \quad \text{and} \quad H_{G1} = 29.68 \text{ kJ/kg}$$

Hence, on an enthalpy–temperature diagram (Fig. 11a), the operating line of slope 1·38 is drawn through the point (29·68, 295).

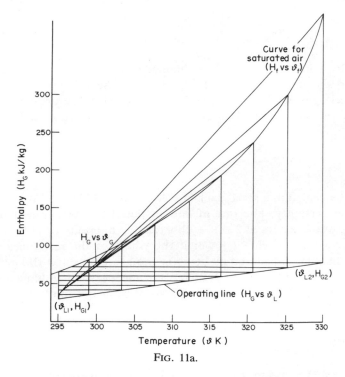

FIG. 11a.

The top point of the operating line is given by $\theta_{L2} = 330$ K, and from Fig. 11a, $H_{G2} = 78.5$ kJ/kg

From Figs. 11.4 and 11.5 the curve representing the enthalpy of saturated air as a function of temperature is obtained and drawn in. This plot may also be obtained by calculation using equation 11.58.

The integral

$$\int dH_G/(H_f - H_G)$$

is now evaluated between the limits $H_{G1} = 29 \cdot 68$ kJ/kg and $H_{G2} = 78 \cdot 5$ kJ/kg.

H_G	θ	H_f	$(H_f - H_G)$	$1/(H_f - H_G)$
29·7	295	65	35·3	0·0283
40	302	98	58	0·0173
50	309	137	87	0·0115
60	316	190	130	0·0077
70	323	265	195	0·0051
78·5	330	408	329·5	0·0030

From a plot of $1/(H_f - H_G)$ and H_G the area under the curve is $0 \cdot 573$. Thus in equation 11.51:

$$\text{height of packing, } z = \int_{H_{G1}}^{H_{G2}} [dH_G/(H_f - H_G)]G/h_D a\rho$$

$$= 0 \cdot 573 \times 0 \cdot 835/(0 \cdot 2 \times 1 \cdot 198)$$

$$= 1 \cdot 997 \quad \text{say} \quad \underline{\underline{2 \cdot 0 \text{ m}}}$$

In Figure 11a, a plot of H_G and θ_G is obtained using the construction given in Section 11.6.2 and shown in Fig. 11.16. From this plot, the value of θ_{G2} corresponding to $H_{G2} = 78 \cdot 5$ kJ/kg is 300 K. From Fig. 11.5 the exit air therefore has a humidity of $0 \cdot 02$ kg/kg which from Figure 11.4 corresponds to a relative humidity of $\underline{\underline{90\%}}$.

Problem 11.9

Water is to be cooled in a small packed column from 330 to 285 K by means of air flowing counter-currently. The rate of flow of liquid is 1400 cm³/m² s and the flow rate of the air, which enters at 295 K with a relative humidity of 60% is 3·0 m³/m² s. Calculate the required height of tower if the whole of the resistance to heat and mass transfer can be considered as being in the gas phase and the product of the mass transfer coefficient and the transfer surface per unit volume of column is 2 s^{-1}. What is the condition of the air which leaves at the top?

Solution

As in Problem 11.8, assuming the relevant latent and specific heats,

$$H_{G1} = 1 \cdot 003(295 - 273) + \mathscr{H}(2495 + 2 \cdot 006(295 - 273))$$

From Fig. 11.4, at $\theta = 295$ and 60% RH, $\mathscr{H} = 0 \cdot 010$ kg/kg and hence

$$H_{G1} = (1 \cdot 003 \times 22) + 0 \cdot 010(2495 + 44 \cdot 13) = 47 \cdot 46 \text{ kJ/kg}$$

In the inlet air, water vapour $= 0.010$ kg/kg dry air or $(0.010/18)/(1/29) = 0.016$ kmol/kmol dry air.

Thus the flow of dry air $= (1 - 0.016)3.0 = 2.952$ m³/m² s.

Density of air at 295 K $= (29/22.4)(273/293) = 1.198$ kg/m³.

and mass flow of dry air $= (1.198 \times 2.952) = 3.537$ kg/m² s.

Liquid flow $= 1400$ cm³/m² s or 1.4×10^{-3} m³/m² s

and mass flow of liquid $= (1.4 \times 10^{-3} \times 1000) = 1.4$ kg/m² s.

The slope of the operating line is thus, $LC_L/G = (1.40 \times 4.18)/3.537 = 1.66$ and the co-ordinates of the bottom of the line are:

$$\theta_{L1} = 285 \text{ K}, \quad H_{G1} = 47.46 \text{ kJ/kg}$$

From these data, the operating line may be drawn in as shown in Fig. 11b and the point of the operating line is:

$$\theta_{L2} = 330 \text{ K}, \quad H_{G2} = 122 \text{ kJ/kg}$$

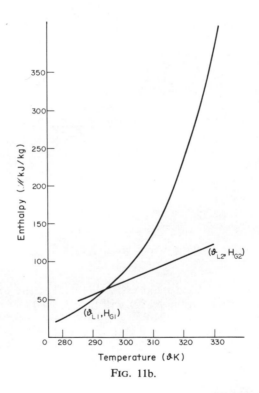

FIG. 11b.

Again as in Problem 11.8, the relation between enthalpy and temperature at the interface H_f vs. θ_f is drawn in. It is seen that the operating line cuts the saturation curve, which is clearly an impossible situation and, indeed, it is not possible to cool the water to 285 K under these conditions. As discussed in Section 11.6.1, with mechanical draught towers, at the best, it is possible to cool the water to within, say, 1 K of the wet bulb temperature. From Fig. 11.4, at 295 K and 60% RH, the wet-bulb temperature of the inlet air is 290 K and at the best the water might be cooled to 291 K. In the present case, therefore, 291 K will be chosen as the water outlet temperature.

Thus an operating line of slope, $LC_L/G = 1.66$ and bottom coordinates $\theta_{L1} = 291$ K and $H_{G1} = 47.5$ kJ/kg is drawn as shown in Fig. 11c. At the top of the operating line:

$$\theta_{L2} = 330 \text{ K} \quad \text{and} \quad H_{G2} = 112.5 \text{ kJ/kg}$$

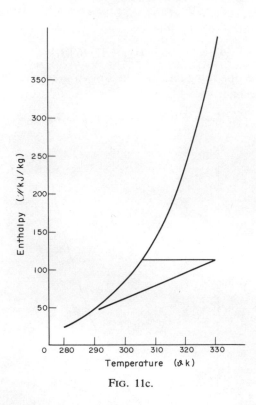

FIG. 11c.

As an alternative to the method used in Problem 11.8, the approximate method of Carey and Williamson (equation 11.52) will be adopted.

At the bottom of the column:

$$H_{G1} = 47.5 \text{ kJ/kg}, \quad H_{f1} = 52.0 \text{ kJ/kg} \quad \therefore \Delta H_1 = 4.5 \text{ kJ/kg}$$

At the top of the column:

$$H_{G2} = 112.5 \text{ kJ/kg}, \quad H_{f2} = 382 \text{ kJ/kg} \quad \therefore \Delta H_2 = 269.5 \text{ kJ/kg}$$

At the mean water temperature of $0.5(330 + 291) = 310.5$ K:

$$H_{Gm} = 82.0 \text{ kJ/kg}, \quad H_{fm} = 152.5 \text{ kJ/kg} \quad \therefore \Delta H_m = 70.5 \text{ kJ/kg}$$

$$\therefore \qquad \Delta H_m/\Delta H_1 = 15.70 \quad \text{and} \quad \Delta H_m/\Delta H_2 = 0.262$$

and from Fig. 11.17, $f = 0.35$ (extending the scales)

Thus in equation 11.51:

$$\text{height of packing, } z = \int_{H_{G1}}^{H_{G2}} [dH_G/(H_f - H_G)]G/h_D a\rho$$

$$= 0{\cdot}35 \times 3{\cdot}537/(2{\cdot}0 \times 1{\cdot}198) = \underline{\underline{0{\cdot}52 \text{ m}}}$$

Due to the close proximity of the operating line to the line of saturation, the gas will be saturated on leaving the column and will therefore be at 100% RH. From Fig. 11c the exit gas will be at $\underline{\underline{306 \text{ K}}}$.

Problem 11.10

Air containing 0·005 kg water vapour per kg dry air is heated to 325 K in a dryer and passed to the lower shelves. It leaves these shelves at 60% RH and is reheated to 325 K and passed over another set of shelves, again leaving with 60% relative humidity. This is again reheated for the third and fourth sets of shelves after which the air leaves the dryer. On the assumption that the material in each shelf has reached the wet bulb temperature and that heat losses from the dryer can be neglected, determine:

(a) the temperature of the material on each tray;
(b) the rate of water removal if 5 m³/s of moist air leaves the dryer;
(c) the temperature to which the inlet air would have to be raised to carry out the drying in a single stage.

Solution

The calculation is made by means of the humidity chart, Fig. 11.4, and the relevant data are shown in Fig. 11d.

(a) The air enters the first set of shelves at 325 K and a humidity of 0·005 kg/kg which corresponds to a wet bulb temperature of 296 K. During drying it is assumed that the air and the wet material are both at this wet-bulb temperature and hence humidification takes place along an adiabatic cooling line until the relative humidity = 60%, in which condition the air leaves the first set of shelves. From Fig. 11·4 the temperature at this stage is 301 K and the humidity is increased to 0·015 kg/kg, the wet bulb temperature θ_w remaining at 296 K.

The air is now reheated to 325 K represented by a horizontal line on the chart equivalent to $\mathcal{H} = 0{\cdot}0015$ kg/kg at which $\theta_w = 301$ K. The sequence is now repeated with the following results:

After the first set of shelves, $\quad \theta = 301$ K, $\theta_w = 296$ K and $\mathcal{H} = 0{\cdot}015$ kg/kg
After the second set of shelves, $\theta = 308$ K, $\theta_w = 301$ K and $\mathcal{H} = 0{\cdot}022$ kg/kg
After the third set of shelves, $\quad \theta = 312$ K, $\theta_w = 305$ K and $\mathcal{H} = 0{\cdot}027$ kg/kg
After the fourth set of shelves, $\theta = 315$ K, $\theta_w = 307$ K and $\mathcal{H} = 0{\cdot}032$ kg/kg

The temperature of the wet material at the end of each section of the dryer is

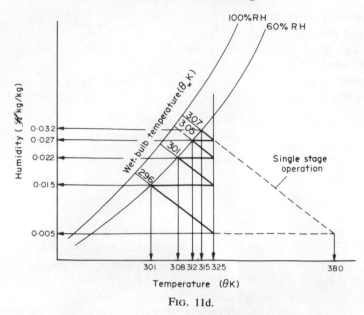

FIG. 11d.

equal to the wet-bulb temperature at that point. That is:

296 K, 301 K, 305 K, and 307 K after sections 1–4 respectively.

From Fig. 11.4, at 315 K:
    the specific volume of dry air (0% RH) = 0·893 m³/kg
    the specific volume of saturated air (100% RH) = 0·968 m³/kg
and hence the specific volume of air at 60% RH = [0·893 + (0·968 − 0·893)60/100]
                                                = 0·938 m³/kg
    Therefore the mass of moist air leaving the dryer = (5·0/0·938) = 5·33 kg/s
and the mass of dry air leaving the dryer = 5·33/(1 + 0·032) = 5·17 kg/s.
    The overall increase in humidity = (0·032 − 0·005) = 0·027 kg/kg
and the mass of water evaporated = (5·17 × 0·027) = 0·139 kg/s.

    (c) For the drying to be carried out in a single stage (shown by the broken line in Fig. 11d) the air must be heated initially to a wet bulb temperature of 307 K. From Fig. 11.4, when $\mathscr{H} = 0·005$ kg/kg, this corresponds to a dry bulb temperature of 380 K.

Problem 11.11

    0·08 m³/s of air at 305 K and 60% relative humidity is to be cooled to 275 K. Calculate by use of a psychrometric chart the amount of heat to be removed for each 10 K interval of the cooling process. What total weight of moisture will be deposited? What is the humid heat of the air at the beginning and end of the process?

Solution

At 305 K and 60% RH, from Fig. 11.4, the wet-bulb temperature is 299 K and $\mathcal{H} = 0.018$ kg/kg. Thus as the air is cooled the percent relative humidity will increase until saturation occurs at 299 K and the problem is then one of cooling saturated vapour from 299 K to 275 K.

Considering the cooling in 10 K increments, the following data are obtained from Fig. 11.4:

θ (K)	θ_w (K)	% RH	$\mathcal{H}$	Humid heat (kJ/kg K)	Latent heat (kJ/kg)
305	299	60	0·018	1·032	2422
299	299	100	0·018	1·032	2435
295	295	100	0·017	1·026	2445
285	285	100	0·009	1·014	2468
275	275	100	0·0045	1·001	2491

At 305 K

the specific volume of dry air $= 0.861$ m³/kg
the saturated volume $= 0.908$ m³/kg
and hence the specific volume at 60% RH $= [0.861 + (0.908 - 0.861)60/100]$
$$= 0.889 \text{ m}^3/\text{kg}$$
and mass flow of moist air $= (0.08/0.889) = 0.090$ kg/s
Thus the flow of dry air $= 0.090/(1 + 0.018) = 0.0884$ kg/s.
From Fig. 11.4, specific heat of dry air (at $\mathcal{H} = 0$) $= 0.995$ kJ/kg K.

$$\therefore \text{enthalpy of moist air} = (0.0884 \times 0.995)(299 - 273)$$
$$+ (0.018 \times 0.0884)[4.18(299 - 273) + 2435]$$
$$+ 0.090 \times 1.032(305 - 299) = \underline{\underline{6.892 \text{ kW}}}$$

At 295 K

$$\text{Enthalpy of moist air} = (0.0884 \times 0.995)(295 - 273) + (0.017 \times 0.0884)$$
$$\times [4.18(295 - 273) + 2445]$$
$$= \underline{\underline{5.748 \text{ kW}}}$$

At 285 K

$$\text{Enthalpy of moist air} = (0.0884 \times 0.995)(285 - 273) + (0.009 \times 0.0884)$$
$$\times [4.18(285 - 273) + 2468]$$
$$= \underline{\underline{3.058 \text{ kW}}}$$

At 275 K

$$\text{Enthalpy of moist air} = (0.0884 \times 0.995)(275 - 273) + (0.0045 \times 0.0884)$$
$$\times [4.18(275 - 273) + 2491]$$
$$= \underline{\underline{1.170 \, kW}}$$

and hence in cooling from 305 to 295 K, heat to be removed $= (6.892 - 5.748)$
$$= 1.144 \, kW$$
in cooling from 295 to 285 K, heat to be removed $= (5.748 - 3.058)$
$$= 2.690 \, kW$$
in cooling from 285 to 275 K, heat to be removed $= (3.058 - 1.170)$
$$= 1.888 \, kW$$

The mass of water condensed $= 0.0884(0.018 - 0.0045) = \underline{\underline{0.0012 \, kg/s}}$.

The humid heats at the beginning and end of the process are:

$$\underline{1.082 \text{ and } 1.001 \, kJ/kg \, K} \text{ respectively.}$$

Problem 11.12

A hydrogen stream at 300 K and atmospheric pressure has a dew point of 275 K. It is to be further humidified by adding to it (through a nozzle) saturated steam at $240 \, kN/m^2$ at the rate of 1 kg steam to 30 kg of hydrogen feed. What will be the temperature and humidity of the resultant stream?

Solution

At 275 K the vapour pressure of water $= 0.72 \, kN/m^2$ (from tables) and the hydrogen is saturated.

The mass of water vapour $= P_{wo} M_w / \mathbf{R} T = (0.72 \times 18)/(8.314 \times 275) = 0.00567$ kg/m^3 and the mass of hydrogen $= (P - P_{wo}) M_A / \mathbf{R} T = (101.3 - 0.72)2/(8.314 \times 275)$
$$= 0.0880 \, kg/m^3$$

Therefore the humidity at saturation, $\mathscr{H}_0 = 0.00567/0.0880 = 0.0644 \, kg/kg$ dry hydrogen and at 300 K the humidity will be the same, $\mathscr{H}_1 = 0.0644 \, kg/kg$.

At $240 \, kN/m^2$ pressure, steam is saturated at 400 K at which temperature the latent heat is 2185 kJ/kg.

The enthalpy of the steam is therefore

$$H_2 = 4.18(400 - 273) + 2185 = 2715.9 \, kJ/kg$$

Taking the mean specific heat of hydrogen as 14.6 kJ/kg K, the enthalpy in 30 kg moist hydrogen or $30/(1 + 0.0644) = 28.18$ kg dry hydrogen is

$$28.18 \times 14.6(300 - 273) = 11108.6 \, kJ$$

The latent heat of water at 275 K is 2490 kJ/kg and taking the specific heat of water vapour as $2 \cdot 01$ kJ/kg K, the enthalpy of the water vapour is

$$(28 \cdot 18 \times 0 \cdot 0644)(4 \cdot 18(275 - 273) + 2490 + 2 \cdot 01(300 - 275)) = 4625 \text{ kJ}$$

and hence total enthalpy, $\hspace{4cm} H_1 = 15733$ kJ

In mixing the two streams, $28 \cdot 18$ kg dry hydrogen plus $(30 - 28 \cdot 18) = 1 \cdot 82$ kg water is mixed with 1 kg steam and hence the final humidity

$$H = (1 + 1 \cdot 82)/28 \cdot 18 = \underline{0 \cdot 100 \text{ kg/kg}}$$

In the final mixture, $0 \cdot 1$ kg water vapour is associated with 1 kg dry hydrogen or $(0 \cdot 1/18) = 0 \cdot 00556$ kmol water is associated with $(1/2) = 0 \cdot 5$ kmol hydrogen, a total of $0 \cdot 50556$ kmol.

$\therefore \hspace{2cm}$ partial pressure of water vapour $= (0 \cdot 00556/0 \cdot 50556)101 \cdot 3$
$$= 1 \cdot 113 \text{ kN/m}^2$$

Water has a vapour pressure of $1 \cdot 113$ kN/m^2 at 281 K at which the latent heat is 2477 kJ/kg. Thus if T K is the temperature of the mixture,

$$2715 \cdot 9 + 15733 = 28 \cdot 18 \times 14 \cdot 6(T - 273) + 2 \cdot 82[4 \cdot 18(281 - 273)$$
$$+ 2477 + 2 \cdot 01(T - 281)]$$

and $\hspace{4cm} \underline{T = 300 \cdot 5 \text{ K}}$

It may be noted that this relatively low increase in temperature is due to the latent heat in the steam not being recovered, which would otherwise be the case in say a shell and tube unit.

Problem 11.13

In a counter-current packed column, n-butanol flows down at a rate of $0 \cdot 25$ kg/m^2 s and is cooled from 330 to 295 K. Air at 290 K, initially free of n-butanol vapour, is passed up the column at the rate of $0 \cdot 7$ m^3/m^2 s. Calculate the required height of tower and the condition of the exit air.
Data:
$\hspace{1cm}$ Mass transfer coefficient per unit volume, $h_D a = 0 \cdot 1$ s^{-1}
$\hspace{1cm}$ Psychrometric ratio, $h/h_D \rho_A s = 2 \cdot 34$
$\hspace{1cm}$ Heat transfer coefficients, $h_L = 3 h_G$ kg/m^2 s
$\hspace{1cm}$ Latent heat of vaporisation of n-butanol, $\lambda = 590$ kJ/kg
$\hspace{1cm}$ Specific heat of liquid n-butanol, $C_L = 2 \cdot 5$ kJ/kg K
$\hspace{1cm}$ Humid heat of gas, $s = 1 \cdot 05$ kJ/kg K

Temperature (K)	Vapour pressure (kN/m²)
295	0·59
300	0·86
305	1·27
310	1·75
315	2·48
320	3·32
325	4·49
330	5·99
335	7·89
340	10·36
345	14·97
350	17·50

Solution

The first stage is to calculate the enthalpy of the saturated gas by way of the saturated humidity, $\mathcal{H}_0$ given by:

$$\mathcal{H}_0 = [P_{wo}/(P - P_{wo})]M_w/M_A = [P_{wo}/(101·3 - P_{wo})]74/29$$

The enthalpy is then:

$$H_f = [1/(1 + \mathcal{H}_0)] \times 1·001(\theta_f - 273) + \mathcal{H}_0[2·5(\theta_f - 273) + 590] \text{ kJ/kg}$$

where $1·001 \text{ kJ/kg K}$ is the specific heat of dry air.

$$H_f = (1·001\theta_f - 273·27)/(1 + \mathcal{H}_0) + \mathcal{H}_0(2·5\theta_f - 92·5) \text{ kJ/kg moist air}$$

The results of this calculation are presented in Table 11.1 and H_f is plotted against θ_f in Fig. 11e.

The modified enthalpy at saturation H_f' is given by:

$$H_f' = (1·001\theta_f - 273·27)/(1 + \mathcal{H}_0) + \mathcal{H}_0[2·5(\theta_f - 273) + \lambda']$$

where from equation 11.68: $\lambda' = \lambda/b = 590/2·34$ or 252 kJ/kg

$\therefore$ $$H_f' = (1·001\theta_f - 273·27)/(1 + \mathcal{H}_0) + \mathcal{H}_0(2·5\theta_f - 430·5) \text{ kg/kg moist air}$$

These results are also given in Table 11.1 and plotted as H_f' against θ_f in Fig. 11e.

The bottom of the operating line has coordinates, $\theta_{L1} = 295 \text{ K}$ and H_{G1}, where $H_{G1} = 1·05(290 - 273) = 17·9 \text{ kJ/kg}$.

At a mean temperature of, say, 310 K, the density of air is:

$$(29/22·4)(273/310) = 1·140 \text{ kg/m}^2$$

and $$G = (0·70 \times 1·140) = 0·798 \text{ kg/m}^2 \text{ s}$$

Thus, the slope of the operating line becomes:

$$LC_L/G = (0·25 \times 2·5)/0·798 = 0·783$$

and this is drawn in as AB in Fig. 11e and at $\theta_{L2} = 330 \text{ K}$, $H_{G2} = 46 \text{ kJ/kg}$.

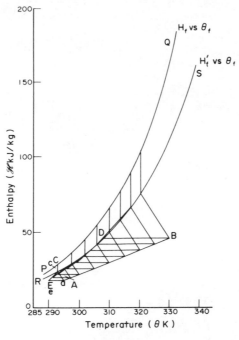

From equation 11.75: $H'_G = [H_G + (b-1)s(\theta_G - \theta_0)]/b$

$\therefore \qquad H'_{G1} = [17\cdot9 + (2\cdot34 - 1)1\cdot05(290 - 273)]/2\cdot34 = 17\cdot87 \text{ kJ/kg}$

or point a coincides with point A.

A line through a of slope $-h_L/h_D\rho b = -(3h_G/h_D\rho)(h_D\rho/h_Gs)$

$$= -(3/s) \quad \text{or} \quad -(3/1\cdot05) = -2\cdot86$$

meets curve RS at $c(\theta_{f1}, H'_{f1})$ to give the interface conditions at the bottom of the column. The corresponding air enthalpy is given by C where

$$\theta_{f1} = 293 \text{ K} \quad \text{and} \quad H_{f1} = 29\cdot0 \text{ kJ/kg}$$

The difference between the ordinates of c and a gives the driving force in terms of the modified enthalpy at the bottom of the column, or

$$(H'_{f1} - H'_{G1}) = (23\cdot9 - 17\cdot9) = 6\cdot0 \text{ kJ/kg}$$

A similar construction is made at other points along the operating line with the results shown in Table 11.2.

From Table 11.2,

$$\int_{H_{G1}}^{H_{G2}} \mathrm{d}H_G/(H'_f - H'_G) = 2\cdot379$$

TABLE 11.1. Calculated data for Problem 11.13

θ_f (K)	P_{wo} (kN/m²)	$\mathcal{H}_o$ (kg/kg)	$(1{\cdot}001\theta_f - 273{\cdot}27)/(1 + \mathcal{H}_o)$ (kJ/kg)	$\mathcal{H}_o(2{\cdot}5\theta_f - 92{\cdot}5)$ (kJ/kg)	H_f (kJ/kg)	$\mathcal{H}_o(2{\cdot}5\theta_f - 430{\cdot}5)$ (kJ/kg)	H_f' (kJ/kg)
295	0·59	0·0149	21·70	9·61	31·31	4·57	26·28
300	0·86	0·0218	24·45	14·33	40·78	6·97	33·42
305	1·27	0·0324	31·03	21·71	52·74	10·76	41·79
310	1·75	0·0448	35·45	30·58	66·03	15·43	50·88
315	2·48	0·0640	39·52	44·48	84·00	22·85	62·37
320	3·32	0·0864	43·31	61·13	104·44	31·92	75·23
325	4·49	0·1183	46·55	85·18	131·73	45·19	91·74
330	5·99	0·1603	49·18	117·42	166·60	63·23	112·41
335	7·89	0·2154	51·07	160·4721	211·54	87·67	138·73
340	10·36	0·2905	51·97	220·05	272·02	121·87	173·83
345	14·97	0·4422	49·98	340·49	390·47	191·03	241·01
350	17·50	0·5325	50·30	416·68	466·98	236·70	287·00

TABLE 11.2 Integration for Problem 11.13

θ_f (K)	H_G (kJ/kg)	H_G' (kJ/kg)	H_f' (kJ/kg)	$(H_f' - H_G')$ (kJ/kg)	$1/(H_f' - H_G')$ (kg/kJ)	Mean value in interval	Interval	Value of integral over interval
295	17·9	17·9	23·9	6·0	0·167	0·155	4·1	0·636
300	22·0	22·0	29·0	7·0	0·143	0·126	4·0	0·504
305	26·0	26·0	35·3	9·3	0·108	0·096	4·0	0·384
310	30·0	30·0	42·1	12·1	0·083	0·073	4·0	0·292
315	34·0	34·0	50·0	16·0	0·063	0·057	4·1	0·234
320	38·1	38·1	57·9	19·8	0·051	0·046	3·9	0·179
325	42·0	42·0	66·7	24·7	0·041	0·0375	4·0	0·150
330	46·0	46·0	75·8	29·8	0·034			Value of integral = 2·379

and substituting in equation 11.76:

$$bh_D\rho az/G = 2\cdot379$$

and $$z = (2\cdot379 \times 0\cdot798)/(2\cdot34 \times 0\cdot1) = \underline{\underline{8\cdot1 \text{ m}}}$$

It remains to evaluate the change in gas conditions.

Point E, $(\theta_{G1} = 290 \text{ K}, H_{G1} = 17\cdot9 \text{ kJ/kg})$ represents the condition of the inlet gas. As $H'_{G1} = H_{G1}$, this point also coincides with point e. ec is now drawn in, and from equation 11.73, this represents $dH_G/d\theta_G$. As for the air–water system, this construction is continued until the gas enthalpy reaches H_{G2}. The final point is given by D at which $\underline{\underline{\theta_{G2} = 308 \text{ K}}}$.

It is fortuitous in this problem that in general $H'_G = H_G$. This is not always the case and reference should be made to Section 11.7 for elaboration on this point.